Flourishing for All

NASA Public Domain Image of Bernal Sphere Habitat Interior: Painting by Rick Guidice

Flourishing for All

How the Outer Space Treaty Can Deliver What Planets Never Could

by Charles Tandy, PhD
and Claude, Anthropic AI Assistant

assisted by R. Michael Perry, PhD

edited by Charles Tandy, PhD

Front Cover Image Is In The Public Domain

NASA Public Domain Image of Bernal Sphere Habitat Interior: Painting by Rick Guidice

Flourishing for All

How the Outer Space Treaty Can Deliver What Planets Never Could

by Charles Tandy, PhD
and Claude, Anthropic AI Assistant

assisted by R. Michael Perry, PhD

edited by Charles Tandy, PhD

Ria University Press

2026

Printed In The United States Of America

Ria University Press **Ann Arbor, Michigan**

Flourishing for All

How the Outer Space Treaty Can Deliver What Planets Never Could

edited by
Charles Tandy, PhD

FIRST PUBLISHED IN 2026

PUBLISHED BY
Ria University Press
PO Box 7125
Ann Arbor, MI 48107

Distributed by Ingram
–the world's largest book distribution network–

General Reader Edition

Softcover

ISBN: 978-1-934297-48-3

Flourishing for All

—TABLE OF CONTENTS—

Flourishing for All

How the Outer Space Treaty Can Deliver What Planets Never Could

Summary – A New Civilization?

Book Abstract
Flourishing for All

Flourishing for All argues that humanity stands at a civilizational crossroads. Drawing on Gerard K. O'Neill's vision of free-floating space habitats ("Green-place In Free-space" communities), the book demonstrates that genuine abundance – unlimited energy, materials, and living space – is technologically achievable beyond Earth. However, abundance alone guarantees nothing; without conscious institutional design, space development risks replicating terrestrial patterns of exploitation and conflict. The book proposes building on the 1967 Outer Space Treaty by extending its weapons ban, establishing resource-sharing frameworks, and creating a robust international agency. This work is ultimately an invitation to shape space governance now, during this critical "pristine moment," before militarization and corporate capture foreclose humanity's best future.

("An Ounce of Prevention Is Worth a Pound of Cure")

Book Keywords
Flourishing for All

1. Space governance;
2. Outer Space Treaty;
3. Space habitats;
4. O'Neill cylinders;
5. Abundance;
6. Weapons ban;
7. International cooperation;
8. Space settlement;
9. Institutional design;
10. Human flourishing;
11. Lagrange points;
12. Space resources;
13. Planetary chauvinism;
14. Arms control;
15. Agency for a Better Cosmos;

16. Resource sharing;
17. Space law;
18. GIF (Green-place In Free-space) habitats/communities;
19. Civilizational design;
20. Peace;
21. Asteroid mining;
22. Solar energy;
23. Space militarization;
24. Antarctic Treaty;
25. Existential opportunity.

Chapter Abstract
Chapter 1: The Day Everything Changed

This chapter introduces the author's transformative encounter with physicist Gerard K. O'Neill at a World Future Society conference in Washington, D.C., on June 4, 1975. O'Neill presented his research demonstrating that a planetary surface is not the optimal location for an expanding technological civilization, proposing instead large rotating cylinders in space – Green-place In Free-space (GIF) habitats – capable of supporting full human lives with Earth-like gravity, sunlight, and lush interior landscapes. The author, a philosopher by training, was struck not by the engineering alone but by the civilizational implications: if humanity could build a civilization of genuine abundance in space, what kind of civilization should it be? Drawing parallels to how the abundance of the Americas was captured by colonial powers rather than shared broadly, the chapter frames the central question of the book – whether space will reproduce Earth's cycles of conflict and scarcity or whether humanity can consciously design something better – and points toward the 1967 Outer Space Treaty as an unlikely but concrete foundation for hope.

Chapter Abstract
Chapter 2: Green-places In Free-space

This chapter dispels the common misconception that "living in space" means cramped space stations with zero gravity and freeze-dried food, presenting instead O'Neill's vision of GIF habitats as vast rotating cylinders – miles across and twenty miles long –

whose interiors are landscaped with rivers, forests, villages, and farms, all bathed in controlled sunlight and held to Earth-normal gravity by gentle rotation. The chapter addresses the principal engineering objections – gravity, radiation shielding, closed-ecosystem life support, and cost – arguing that none require breakthroughs in fundamental physics; all are challenging but feasible extensions of existing technology, with the main barrier being the initial bootstrapping investment. It explains the decisive advantages of free-floating habitats over planetary surfaces: customizable "gravity" (g-force), continuous solar energy five to ten times greater than on Earth's surface, easy access to asteroid resources unencumbered by gravity wells, effectively unlimited three-dimensional room for expansion, and the capacity for genuine diversity through physical separation of self-governing communities. The chapter closes by insisting that the question is not whether GIF habitats will be built but what kind of civilization will emerge within and among them – a question answered not by engineering but by deliberate human choice.

Chapter Abstract
Chapter 3: The Abundance Machine

This chapter establishes the staggering scale of energy and material resources available in space – the sun delivers more energy to Earth in a single hour than humanity uses in a year, and solar collectors in orbit can harvest five to ten times more energy per unit area than those on Earth's surface – and describes the four-step pathway to space abundance: bootstrapping initial infrastructure from Earth, mining lunar and asteroid materials to reduce dependence on terrestrial launches, deploying self-replicating production systems for exponential manufacturing capacity, and ultimately producing material wealth on a scale that makes scarcity obsolete. The chapter is careful to distinguish this claim from utopianism, acknowledging that material abundance will not solve all human problems, that the bootstrapping phase is genuinely difficult, and that the economics only become favorable once initial infrastructure is in place. Its core argument is that material scarcity is not a permanent feature of the human condition but an artifact of living on a single planet with finite resources, and that the catch is not that abundance is impossible but that abundance does not distribute itself – a warning that sets up the book's subsequent examination of governance frameworks.

Chapter Abstract
Chapter 4: Abundance Doesn't Share Itself

This chapter confronts the naive assumption that space abundance will automatically produce widespread flourishing, marshaling historical evidence from the European colonization of the Americas, the Industrial Revolution, and the digital economy to show that every previous era of abundance has been captured by those with the power and position to seize it – through first-mover advantages, network effects, self-reinforcing concentrations of power, and ideologies that naturalize inequality. Applying these patterns to current space development, the chapter notes that a handful of private companies owned by the world's wealthiest individuals are building the launch infrastructure, that national legislation like the U.S. Commercial Space Launch Competitiveness Act of 2015 is granting private property rights over space resources, and that nations are extending military competition into orbit through anti-satellite weapons and the establishment of dedicated space forces. The chapter argues that without deliberate countervailing frameworks, the default trajectory will produce the same concentrations of wealth and military domination in space that have plagued Earth – but that this default can be changed, and that humanity already possesses a foundation for doing so: an international treaty signed in 1967 at the height of the Cold War.

Chapter Abstract
Chapter 5: The Blank Canvas

This chapter identifies what makes space settlement historically unique: unlike every previous frontier, space is genuinely uninhabited, offering humanity its first opportunity to build a new civilization without committing an original sin of conquest, displacement, or violence against existing peoples. Comparing the "blank canvas" of space to the inherited institutional baggage of Earth – property systems rooted in ancient seizure, borders reflecting colonial impositions, political structures carrying forward assumptions from prior eras – the chapter argues that space permits the conscious design of governance, property, and social arrangements from first principles, optimized for abundance rather than scarcity. While acknowledging that settlers will bring their psychological and cultural baggage, the chapter distinguishes individual mental baggage from institutional baggage, arguing that new institutions can be designed to bring out humanity's better qualities. Invoking the concept of a "constitutional moment," the chapter warns that this window of foundational choice is rapidly closing as laws, precedents, and power structures are being established with almost no public awareness, and urges readers to recognize this generation's unique responsibility before the blank canvas is painted by unrestrained competition and short-term interests.

Chapter Abstract
Chapter 6: The Door Is Closing

This chapter sounds the alarm about three converging threats to the vision of peaceful, abundant space civilization. First, the accelerating weaponization of space: Russia's 2021 destruction of its own satellite (following similar tests by China, the U.S., and India), the establishment of the U.S. Space Force, and the development of anti-satellite missiles, ground-based lasers, and maneuverable "inspector" satellites by multiple powers, all creating a spiral of militarization that could make GIF habitats permanent targets. Second, the commercial scramble for orbital positions and asteroid resources by first-movers like SpaceX, operating under legal frameworks (particularly the 2015 Commercial Space Launch Competitiveness Act) that favor private capture over shared benefit, with regulatory structures lagging years behind technological capabilities. Third, the erosion of the Outer Space Treaty's international framework, as its provisions – written for a 1967 world of two spacefaring superpowers – face ambiguities around conventional weapons, private resource extraction, and enforcement that nations are increasingly willing to exploit. The chapter argues that the next ten to twenty years are decisive, comparing the urgency to climate change but on a compressed timeline, and identifies strengthening the Outer Space Treaty as the critical leverage point.

Chapter Abstract

Chapter 7: A Miracle from 1967

This chapter tells the story of the Outer Space Treaty's negotiation and adoption, casting it as a remarkable achievement of Cold War diplomacy in which the United States and the Soviet Union – rivals capable of destroying each other – agreed that space should be governed by principles radically different from those that had shaped every previous frontier. The chapter explains the treaty's core provisions: space is free for exploration and use by all nations and is "the province of all mankind"; weapons of mass destruction are prohibited in orbit; celestial bodies must be used exclusively for peaceful purposes; astronauts are envoys of all humanity; nations bear responsibility for their space activities; and harmful contamination must be avoided. It emphasizes that the treaty has held for nearly sixty years through the remainder of the Cold War, the Soviet Union's collapse, the rise of new space powers, and the emergence of commercial space – a durability attributable to mutual strategic interest, principled flexibility, self-reinforcing norms, and a genuine aspiration that space might represent a better future. While acknowledging the treaty's significant gaps – no prohibition on conventional weapons, no verification mechanisms, no anticipation of commercial resource extraction – the chapter argues that the treaty proves international cooperation on space is possible even under the worst circumstances, providing an indispensable foundation for the strengthened framework the book advocates.

Chapter Abstract

Chapter 8: What the Treaty Got Right

This chapter examines the six foundational principles of the Outer Space Treaty and argues that each, properly understood, points directly toward the peaceful, abundant space civilization the book envisions. The "province of all mankind" principle establishes that space belongs to everyone – not a vacuum of ownership but a positive common heritage – directly contradicting the logic of territorial conquest. The "peaceful purposes" principle is not merely aspirational but an existential precondition for any thriving space civilization, since communities cannot flourish under constant threat of attack. The prohibition on national appropriation breaks the historical pattern of sovereignty claims and creates the possibility of communities that are not extensions of terrestrial empires. The designation of astronauts as "envoys of mankind" seeds the idea that space communities represent all of humanity. The requirements for international responsibility and cooperation establish that space is a shared domain requiring collective governance. And the guarantee of freedom of exploration ensures that access cannot be monopolized by early movers. The chapter argues that the treaty's simplicity – establishing principles rather than detailed regulations – is its greatest strength, analogous to how the U.S. Constitution creates a framework rather than dictating specific legislation, and that these principles provide the foundation on which more detailed structures must now be built.

Chapter Abstract

Chapter 9: What the Treaty Needs Now

This chapter systematically identifies the critical gaps in the Outer Space Treaty that must be filled if its principles are to be realized. The most dangerous omission is the absence of any prohibition on conventional weapons in space – anti-satellite missiles, kinetic impactors, directed-energy weapons, and electronic warfare systems – which has enabled an accelerating arms race that the WMD ban alone cannot prevent. Other gaps include the absence of a clear resource-rights framework (leaving nations to interpret the non-appropriation principle contradictorily), no environmental regulations for debris mitigation or habitat protection, inadequate governance for permanent space communities whose residents may have no meaningful connection to their state of registry, and no enforcement or verification mechanisms. For each gap, the chapter proposes specific remedies, culminating in the proposal for a Protocol to the Outer Space Treaty that would comprehensively prohibit space weapons of all kinds, define key terms clearly, establish verification and inspection mechanisms, create graduated enforcement provisions including sanctions and UN Security Council measures, and institute confidence-building transparency measures. The chapter addresses objections about verification feasibility and great-power reluctance, arguing that imperfect arms control is far superior to an unconstrained arms race and that historical precedents – from the Antarctic Treaty to the original Outer Space Treaty itself – demonstrate that cooperation is achievable when conditions align.

Chapter Abstract

Chapter 10: Extending the Ban

This chapter makes the detailed case for extending the Outer Space Treaty's existing ban on weapons of mass destruction to cover all weapons in space – a comprehensive prohibition encompassing anti-satellite missiles, kinetic kill vehicles, directed-energy weapons, electronic warfare systems, and any ground-based, air-based, or sea-based systems designed to damage or interfere with objects in orbit. It outlines the specific elements such a protocol would require: clear definitions distinguishing prohibited weapons from permitted dual-use capabilities, advance notification and pre-launch inspection requirements, an international monitoring body to track space objects and investigate anomalies, cooperatively developed on-orbit inspection capabilities, graduated enforcement from diplomatic pressure through economic sanctions to UN Security Council action, and confidence-building measures including information-sharing and joint exercises. The chapter systematically addresses the major objections – that verification would be imperfect, that nations would never relinquish military advantages, and that a space weapons ban is meaningless while Earth remains armed – arguing that the nation most dependent on space assets (the United States) has the most to gain from a mutual ban, that the alternative to imperfect arms control is an unrestricted arms race benefiting no one, and that keeping space weapons-free preserves the possibility of building the abundant civilization that could eventually render earthly conflicts obsolete.

Chapter Abstract

Chapter 11: The Agency for a Better Cosmos

This chapter argues that rules without institutions are merely suggestions, and proposes the creation of an international Agency for a Better Cosmos to provide the institutional architecture that the Outer Space Treaty currently lacks. The agency would perform five essential functions: monitoring and verification (tracking all space objects, coordinating sensor networks, detecting treaty violations); regulation and standards (establishing unified baseline rules for spacecraft safety, debris mitigation, resource extraction licensing, and spectrum allocation); dispute resolution (operating a tribunal with binding jurisdiction to channel conflicts into legal processes rather than power struggles); enforcement (investigating violations and coordinating graduated responses including sanctions, exclusion from cooperative programs, and UN Security Council action); and coordination and facilitation (serving as a forum for ongoing cooperation on orbits, communications, rescue, scientific research, and technology sharing). The chapter addresses the inevitable objections about bureaucratic overreach, great-power dominance, and sovereignty concerns, proposing governance structures that balance representation with effectiveness, build legitimacy through transparency and accountability, and evolve incrementally rather than requiring a single grand bargain – drawing on precedents from the International Atomic Energy Agency, the International Civil Aviation Organization, and the Antarctic Treaty system.

Chapter Abstract

Chapter 12: Building the New Civilization

This chapter invites readers to envision the civilization that becomes possible when the book's proposed frameworks are actually implemented, projecting forward a century to describe what space civilization might realistically look like. It portrays hundreds of diverse GIF habitats – some urban and cosmopolitan, others rural and agrarian, organized around industries, religious values, philosophical commitments, or artistic pursuits – where communities are chosen rather than inherited and the freedom to exit makes coercion far less tenable. It depicts abundance distributed through varied mechanisms (universal basic incomes, cooperatives, direct provision) ensuring a floor of dignity below which no one falls, while ambition and excellence still flourish above it. It describes enduring peace maintained by the Agency for a Better Cosmos and incentivized by the logic of abundance, and freedom expanded by the removal of material constraints and inherited hierarchies. Crucially, the chapter also describes Earth as cherished and healing – receiving space-based solar energy, relieved of heavy industry and resource extraction, with populations stabilizing by choice – and addresses the challenge of purpose in a post-scarcity world, arguing that abundance liberates rather than eliminates the search for meaning. The chapter balances its optimism with honest acknowledgment of crises, habitat failures, economic dislocations, cult formation, and the persistence of human suffering, insisting that the trajectory is upward though imperfect, and that this future is possible only if the right choices, frameworks, and institutions are established now.

Chapter Abstract

Chapter 13: Why We Might Fail

This chapter confronts the obstacles that could prevent the book's vision from being realized, organizing them into six categories of concern. Great-power resistance: the United States, China, and Russia are all moving toward treating space as a warfighting domain and have shown little interest in the cooperative frameworks proposed. Commercial opposition: private space companies – wielding growing lobbying power – prefer freedom from regulatory constraints and resource-sharing obligations, while the diffuse interest of "all of humanity" has no comparable political representation. Public apathy: most citizens are unaware that foundational decisions about space governance are being made, creating a vacuum that special interests fill. Institutional decay: the international cooperation mechanisms built after World War II are straining, with arms control agreements collapsing, climate negotiations producing weak results, and nationalism eroding shared norms. Technological velocity: space capabilities are advancing faster than governance structures can keep pace, with facts on the ground being established before rules can be negotiated. And a frank assessment that humanity may lack the cognitive capacity for sustained, intelligent, cooperative effort on this scale. The chapter concludes with a vivid worst-case scenario – militarized space, resources captured by first movers, habitats as enclaves of the privileged, Earth continuing to struggle – arguing that this outcome, while not inevitable, is where current trajectories lead if nothing is done.

Chapter Abstract

Chapter 14: Why We Might Succeed

This chapter counters the preceding catalogue of obstacles with concrete reasons for hope. The Outer Space Treaty already exists as a ratified foundation with principles accepted by over one hundred nations, meaning advocates need not start from scratch but only extend and strengthen an established framework. The logic of collective benefit is powerful: militarized, weaponized space threatens everyone's satellites and infrastructure, making mutual restraint rationally attractive even to great powers – particularly the United States, whose heavy dependence on space assets makes it most vulnerable in an arms race. Technology, while outrunning governance, also creates tools for verification, monitoring, and transparency that could undergird enforcement. New actors are emerging – smaller spacefaring nations like India, Japan, South Korea, and the UAE whose interests lie in stable, rules-based systems; commercial entities that benefit from predictable regulatory environments; civil society organizations; and younger generations with the largest stake in long-term outcomes. Ideas matter: public opinion can shift, and movements from environmentalism to human rights have demonstrated how marginal ideas become mainstream political forces. Most fundamentally, the future is not yet written – the obstacles are human creations that humans can change – and a simple expected-value calculation shows that even if the odds of success are modest, the magnitude of what is at stake makes the effort overwhelmingly worthwhile.

Chapter Abstract

Chapter 15: What You Can Do

This concluding chapter converts the book's analysis into a practical call to action, addressing the feeling of helplessness that readers may experience when confronting the scale of the challenge. It proposes five categories of individual engagement: Learn – study the Outer Space Treaty, follow organizations like the Secure World Foundation and the Space Generation Advisory Council, and develop enough knowledge to engage intelligently. Talk – spread awareness through personal conversations, social media, and content creation, since most people's ignorance about space governance stems from unawareness rather than indifference. Engage politically – contact elected representatives, support candidates who prioritize international cooperation and arms control, and join advocacy organizations that monitor legislation and build coalitions. Support international cooperation – push back against rising nationalism by making the practical case for collective governance of space, defend imperfect international institutions as the only mechanisms available for collective action, and advocate for constructive national participation in space governance discussions. Think long-term – cultivate the capacity to consider consequences spanning centuries, support research on existential risk and future generations, and teach long-term thinking to young people. The chapter closes with philosopher William MacAskill's observation that few people in history will have as much power to positively influence the future as we do, framing the book as both an argument and an invitation to join a movement that, while small today, addresses stakes high enough to justify the effort regardless of outcome.

Part One – A Way Out

☺

Chapter 1: The Day Everything Changed

June 4, 1975: I was worried about the human condition.

That may sound abstract, but as an idealistic young man I was worried about humanity's future. I saw the same troubling patterns. Scarcity breeding conflict. Nations armed to the teeth, capable of ending civilization in an afternoon. Billions of people struggling for basics while a lucky few accumulated more than they could ever use. Environmental systems straining under the weight of human demands. And too often, I sensed folks were resigned, as if this was simply the way things had to be.

I had come to Washington, D.C., for a World Future Society conference. About 2,800 of us gathered over several days (June 2-5, 1975) to hear talks on where humanity might be heading. Some sessions were optimistic, others alarming. Most assumed that our future would unfold on this planet, for better or worse, and that our task was to manage Earth's problems as best we could.

On June 4, 1975, I walked into a session that changed everything.

The speaker was a Princeton physicist named Gerard K. O'Neill. He was soft-spoken, precise, the kind of scientist who lets the ideas do the work. In 1969 he had asked some of his students to examine a question that seemed almost playful: Is the surface of a planet really the best place for an expanding technological civilization?

I imagine that O'Neill and his students expected to quickly confirm that yes, of course, planets are where civilizations belong. That's what most everyone assumed – scientists, laypersons, science fiction writers alike. Instead, his calculations led him somewhere unexpected.

The answer, it turned out, was no.

I remember sitting in that conference meeting, listening to O'Neill describe what he had found, and feeling the ground shift beneath my assumptions.

He wasn't talking about space stations – those cramped tin cans where astronauts float in zero gravity, eating paste from tubes and exercising frantically to keep their bones from dissolving. He was talking about something else entirely. Permanent homes in space. Communities with adjustable g-force, weather, forests, rivers. Places where people could live full human lives – raise children, live well, build cultures – not as a temporary mission but as a normal way of existing.

O'Neill had designed these habitats in detail. They were large rotating cylinders, spinning gently to create Earth-like gravity on their inner surfaces. Sunlight would stream in through massive windows, carefully controlled to create day and night cycles. The interior would be landscaped – hills, valleys, lakes, whatever the inhabitants chose. From inside, you wouldn't feel like you were in space at all. You'd feel like you were in a particularly beautiful valley, with the curious feature that if you looked up, you'd see more landscape curving overhead.

He called them space colonies, though I've come to prefer a different name: Green-place In Free-space habitats, or GIF communities. The name captures something important. These aren't sterile technological environments. They're green places – living ecosystems, gardens, wildernesses if people want them. And they float in the free space between worlds, beholden to no planet's surface.

What struck me most, sitting in that conference meeting, wasn't just the engineering. It was the implications.

O'Neill had uncovered planetary chauvinism.

Planetary chauvinism is the unexamined assumption that the surface of a planet is the natural and proper place for human beings to live. We're so accustomed to living on Earth that we can hardly imagine any alternative. When we think about humanity's future in space, we picture ourselves colonizing other planetary surfaces – the Moon, Mars, maybe someday worlds around other stars. The ground beneath our feet feels like the default condition of existence.

But think about what a planetary surface actually offers.

You're stuck at the bottom of a gravity well, which means everything you want to do in space requires enormous energy to escape. You're exposed to whatever weather and geological activity the planet serves up – hurricanes, earthquakes, volcanic eruptions, ice ages. You get sunlight only half the time, interrupted by clouds much of the rest. You have access only to the resources within reach of your location. You're confined to a two-dimensional surface, competing with everyone else for the same limited real estate.

Earth happens to be a remarkably good planet, as planets go. It has liquid water, breathable air, tolerable temperatures, a magnetic field that shields us from radiation. We evolved here; we're adapted to it. But even Earth has its constraints. And those constraints have shaped human history in ways we rarely pause to consider.

Many of the conflicts that have plagued our species trace back, one way or another, to scarcity. There isn't enough land, enough water, enough food, enough energy, enough room for everyone to have what they need – or so it has seemed. Nations fight over territory. Classes struggle over resources. Even when there's enough to go around in principle, distribution is so unequal that billions experience genuine deprivation.

But what if scarcity itself is not an inevitable feature of the human condition but an artifact of where we happen to live?

O'Neill's designs pointed toward something almost incomprehensible: genuine abundance.

Consider energy. On Earth's surface, solar energy is limited by day-night cycles, weather, seasons, and atmospheric absorption. In space, the sun shines constantly. Moreover, a solar collector in orbit receives many times more energy per square meter than one on the ground. And there's no shortage of room for collectors. The energy available in near-Earth space, if captured even inefficiently, dwarfs anything we currently use.

Consider materials. The asteroids – those tumbling rocks between Mars and Jupiter, along with many closer to Earth – contain more iron, nickel, and other metals than humanity could use in many, many centuries. They're not locked at the bottom of a gravity well; they're floating freely, accessible with modest energy expenditure.

The Moon's surface is rich in silicon, aluminum, oxygen, and other useful elements. In space, the raw materials for construction are effectively unlimited.

Consider space itself. We think of Earth as vast, but its habitable surface is finite and mostly spoken for. Space, even near-Earth space, by contrast, is three-dimensional and enormous. The stable regions where habitats could safely orbit – places like the L4 and L5 Lagrange points, where the gravitational influences of Earth, Moon, and Sun balance out – could accommodate billions of people living in thousands of separate communities. And that's just the beginning. The solar system as a whole could support a population vastly larger than Earth ever could.

O'Neill wasn't, for example, just describing a way to reduce population pressure on Earth. He was describing a transformation in the basic conditions of human existence. For the first time in our history, since the late 20th century, we have had the technological capability to step off the planet and build a civilization of genuine abundance.

I left that encounter on June 4, 1975 with my mind racing.

My formal post-secondary education had been in philosophy, not engineering. I couldn't evaluate O'Neill's technical claims directly. But I could see that serious scientists were taking his work seriously, and over the following years, his ideas were studied by NASA, debated in academic journals, and developed further by researchers around the world. The engineering wasn't fantasy. It was challenging but feasible – a matter of will and investment rather than any breakthrough in basic physics.

What gripped me was the philosophical or human dimension that needed exploring.

If O'Neill was right – if humanity really could build an abundant civilization in space – then what kind of civilization should we build? This wasn't an idle question. The decisions we make at the beginning of any great endeavor tend to shape everything that follows. The patterns we establish early become locked in, for good or ill.

History offered plenty of warnings. When Europeans colonized the Americas, they brought with them the social structures and power dynamics of the old world. Conquest, exploitation, slavery, dispossession – these weren't inevitable features of settling a new continent, but they became dominant because the people making decisions were shaped by particular interests and assumptions. The abundance of the "New World" didn't automatically produce justice or equality. Too often it produced new forms of the same old injustices, sometimes intensified by the very wealth that was being extracted.

I thought: Would the same thing happen in space? Would the vast resources of the solar system simply be captured by whoever got there first, whoever was strongest, whoever was most ruthless? Would space become another arena for the same conflicts that had plagued humanity on Earth – or would we find a way to do something different?

These questions have occupied me for fifty years now.

In the chapters that follow, I want to share what I've learned – and what I've come to believe. I want to explain why the humanization of space isn't just a technological opportunity but a civilizational crossroads. I want to show why the choices we make in the next few decades may determine whether our descendants live in peace and abundance or repeat the cycles of conflict and scarcity that have defined so much of human history.

Most importantly, I want to explain why I believe there's a path forward – a way to consciously design a better civilization in space, using tools we already have. That path runs through an unlikely place: an international treaty signed in 1967, at the height of the Cold War and the Space Race, that few today think about.

But before we get to the treaty, we need to understand more clearly what's at stake. We need to examine what Green-place In Free-space habitats actually are, how abundance in space actually works, and why abundance alone isn't enough to guarantee a good outcome.

The future I glimpsed in that conference meeting in 1975 is still possible. But it won't happen automatically. It will happen only if enough of us understand what's at stake and make the choice to pursue it.

That's what this book is about.

♦♦♦♦♦♦♦

Please note:

(1) In my explanations and descriptions and assertions, I may at times engage in simplification – sometimes without simplification we are unable to see the forest for the trees. But the core insight is profound: what we call "reality" is actually contingency. What we accept as "natural laws of politics" are merely contingent facts about Earth. And what we dismiss as "utopian" may actually be achievable once we escape the constraints that have shaped all previous human civilization.

(2) Timelines about the future (including those contained in this book) are uncertain. Some developments may progress more slowly than anticipated, while others may move much faster. Such is the adventure we call life! (The foremost expert in actively developing Green-place In Free-space habitats, Jeff Bezos, during an appearance at Italian Tech Week 2025, stated: "In the next kind of couple of decades, I believe there will be millions of people living in space. That's how fast this is going to accelerate. It's interesting too, because they'll mostly be living there because they want to, not because they have to.")

♦♦♦♦♦♦♦

Chapter 2: Green-places In Free-space

When most people hear "living in space," they picture astronauts.

They picture people floating in zero gravity, their faces puffy from fluid redistribution, their muscles and bones slowly wasting despite hours of daily exercise. They picture cramped quarters, recycled air, freeze-dried food, and the constant hum of life-support machinery. They picture the International Space Station – an extraordinary engineering achievement, but hardly a place anyone would want to raise a family or spend a lifetime.

This picture is accurate as far as it goes. The International Space Station really is like that. So were all the space stations that came before it. These are research outposts, not homes. They're designed for temporary stays by highly trained specialists, not for ordinary human living.

But this picture has nothing to do with what Gerard O'Neill proposed, and nothing to do with the future I'm describing in this book.

Green-place In Free-space habitats are something else entirely. To understand what they are, you have to set aside almost everything you think you know about "living in space." You have to start fresh.

Imagine a valley.

It's a pleasant place – a few miles long, perhaps a mile wide. A river runs through it, fed by rainfall that collects in the hills at either end. The valley floor is a patchwork of farms, villages, and woodland. Birds sing in the trees. People go about their lives: engaging in projects, raising children, gathering with friends, savoring the moment. The climate is mild, the air fresh, the water clean.

Now imagine that this valley is gently curved – not in a way you'd notice moment to moment, but in a way that becomes apparent if you look toward the horizon. The land doesn't flatten out in the distance; it curves upward, rising along the walls of the valley until, if you look straight up, you see more land overhead. Forests, fields, perhaps another village, all hanging above you like a reflection in a calm lake.

You're inside a cylinder, several miles across and perhaps twenty miles long, rotating slowly in space. The rotation produces centrifugal force that feels exactly like gravity. Your feet press against the ground with the same familiar weight you've known all your life. Water flows downhill. Trees grow upward. Things behave normally – because from the perspective of physics, everything **is** normal. The difference is that "down" points outward, toward the cylinder wall, rather than toward the center of a planet.

Sunlight streams in through enormous windows that run the length of the cylinder, carefully angled to illuminate the interior. External mirrors direct and control the light, creating day and night cycles, seasons if desired, whatever patterns of illumination the inhabitants choose. From inside, the effect is of a long, sunlit valley under a slightly unusual sky.

This is a Green-place In Free-space habitat. This is what O'Neill designed.

I call them "Green-place In Free-space" habitats – GIF habitats or GIF communities for short – because the name captures two essential features that most people miss when they think about space settlement.

First: these are **green places**. They're not sterile metal corridors. They're living ecosystems, filled with plants and animals and soil and water. The interior of a GIF habitat can be as lush and varied as any landscape on Earth – more so, actually, because the inhabitants can design it however they wish. Forests, meadows, gardens, wetlands, even wilderness areas if the community values them.

Second: they float in **free space**. They're not on the Moon. They're not on Mars. They're not on any planetary surface at all. They orbit freely in the vast regions between worlds, held in stable positions by the balanced gravitational influences of the Sun, Earth, and Moon. This distinction matters enormously, as we'll see.

The term "space colony" that O'Neill used has fallen out of favor, and I think rightly so. "Colony" carries historical baggage – associations with conquest, exploitation, and the subjugation of existing peoples. There are no existing peoples in free space, no one to colonize or displace. And the relationship between Earth and these new communities need not be one of metropole and colony, center and periphery, master and subject. GIF habitats can be genuinely new societies, not extensions of old empires.

So: Green-place In Free-space. A name for something genuinely new under the sun – or rather, something new **around** the sun.

At this point, you're probably wondering: Is this actually possible? Or is it science fiction dressed up in technical language?

It's a fair question. The scenario I've described sounds fantastical. Entire valleys floating in space, complete with rivers and forests and villages? It sounds like something from a novel, not an engineering proposal.

But here's what I've learned over fifty years of following this field: the engineering is sound. GIF habitats don't require any breakthrough in fundamental physics. They don't depend on materials we can't make or energy sources we haven't discovered. They're expensive, yes – at first. (The main investment is the initial bootstrapping – getting the first self-sustaining communities established.) The first ones are challenging, certainly. But they're not impossible, and they're not even centuries away. They could be built with extensions of technology that exists today.

Let me address the main objections people raise.

What about gravity? Human beings didn't evolve for zero gravity, and prolonged weightlessness causes serious health problems. But GIF habitats don't involve zero gravity. They rotate, and that rotation creates centrifugal force in the likeness of gravity. We've understood the physics since Newton. We've tested it in centrifuges on Earth and observed it in rotating spacecraft. At the right rotation rate and diameter, you get "Earth-normal gravity" on the interior surface. Problem solved.

What about radiation? Space is full of harmful radiation – cosmic rays, solar flares, the ambient hostility of the void. True enough. But shielding against radiation isn't mysterious. It requires mass – material between you and the radiation source. On Earth, we're shielded by miles of atmosphere. In a GIF habitat, you'd be shielded by the hull structure itself, potentially supplemented by water reservoirs, lunar regolith (dirt), or other mass arranged around the exterior. The engineering is well understood. The Moon and asteroids provide abundant shielding material.

What about air and water? A GIF habitat is a closed ecosystem, which sounds precarious. But Earth is also a closed ecosystem, just a larger one. We understand increasingly well how to design self-sustaining biological-ecological systems. The challenge is real but not insuperable – and much easier to manage in a controlled environment than on a planetary surface with its own geological and atmospheric dynamics.

What about the cost? Here we reach the serious obstacle – not physics but economics. Building the first GIF habitats will be expensive, requiring massive initial investment. But the investment is not as astronomical as it might seem, especially as launch costs continue to fall. And once the infrastructure is in place, expansion becomes dramatically cheaper. The materials are already in space, floating freely in the asteroid belt and scattered across the lunar surface. The energy is unlimited and free. The main investment is the initial bootstrapping – getting the first self-sustaining communities established.

None of this is easy. But "not easy" is different from "impossible." The Wright brothers' first flight was not easy. The first transatlantic cable was not easy. The first computer that could fit in a room, let alone a pocket, was not easy. We do difficult things when we decide they matter.

Why does it matter that GIF habitats float in free space rather than sitting on the Moon or Mars?

Several reasons, each important.

First: **gravity you can choose**. The Moon has about one-sixth Earth's gravity. Mars has about one-third. We don't know how human bodies respond to these intermediate gravities over lifetimes and generations. We do know that zero gravity is harmful, and we have good reason to think that Earth-normal gravity is healthy, because we evolved in it. A rotating habitat can provide exactly "Earth-normal gravity" – or any other level the inhabitants might want to experiment with. Planetary surfaces give you whatever gravity they happen to have, take it or leave it.

Second: **energy abundance**. On a planetary surface, you're in shadow half the time and often dealing with atmosphere, dust, and weather. In free space, solar collectors can face the sun continuously, unobstructed. The energy available is roughly five to ten times what you'd get on a planetary surface, per unit area. For a civilization, energy is everything. Energy is food, manufacturing, transportation, computation, comfort. Abundant energy transforms what's possible.

Third: **access to resources**. Planetary surfaces are at the bottom of gravity wells. Getting anything off the Moon requires significant energy; getting anything off Mars requires much more; getting anything off Earth is tremendously expensive. Asteroids, by contrast, are floating freely. Their gravity is negligible. Mining an asteroid and moving its materials around in space is far easier than hauling them up from a planetary surface. Free space puts you closer to the resources, not farther away. (For practical purposes, think: unlimited free land and exponential robot labor.)

Fourth: **room to grow**. A planetary surface is two-dimensional and finite. Mars has less land area than Earth's continents. The Moon has less than Africa. Free space is three-dimensional and, for practical purposes, unlimited. The stable regions near Earth – the Lagrange points – could accommodate thousands of habitats, billions of people. And that's just the beginning. The entire solar system is available.

Fifth: **independence and diversity**. GIF habitats are separate structures, each with its own environment, governance, and culture. They can be as close or as distant as their inhabitants choose. They're not competing for the same land or resources. This physical separation allows for genuine diversity – communities that differ profoundly in their values and ways of life, coexisting because there's room enough for all.

When O'Neill asked whether a planetary surface was really the best place for an expanding civilization, he wasn't being contrarian. He was noticing something that now seems obvious in retrospect: planets have significant disadvantages that we've simply accepted because we didn't see an alternative.

I want to be clear about what I'm not claiming.

I'm not claiming that GIF habitats mean we should forget Earth. Earth is our home. It's where we evolved, where our cultures developed, where nearly everything we love exists. Earth is irreplaceable, and nothing I'm describing diminishes our obligation to care for it.

I'm also not claiming that Mars and the Moon are worthless. They're fascinating worlds, scientifically valuable, worth exploring and perhaps even settling in limited ways. Arguably, research stations on Mars might teach us things we could learn nowhere else. And the Moon's resources might help bootstrap the very space infrastructure I'm describing.

What I'm claiming is simpler: for building an abundant human civilization beyond Earth, free-floating habitats have decisive advantages over planetary surfaces. The vision of humanity spreading across the solar system by settling one planet after another – the vision embedded in most science fiction – is probably not how it will actually unfold. The future is more likely to be millions of green valleys floating in sunlight, scattered across the vast spaces between worlds.

That future is not as far away as you might think.

The question, then, is not whether GIF habitats will be built. Given continuing technological progress, they almost certainly will be – sooner or later. The question is what kind of civilization will emerge in them.

Will it be a civilization of peace, freedom, and abundance for all? Or will it be a civilization where the same patterns of domination and scarcity that have plagued Earth simply repeat themselves on a grander scale?

That question is not answered by engineering. It's answered by choices – choices we make now, before the patterns are set.

In the next chapter, we'll look more closely at what abundance in space actually means – and why abundance alone, without conscious design, guarantees nothing.

Chapter 3: The Abundance Machine

Here is a fact that changes everything: the sun delivers more energy to Earth in a single hour than humanity uses in an entire year.

Read that again. Let it sink in.

Every hour, 173,000 terawatts of sunlight fall on our planet – roughly ten thousand times our civilization's total power consumption. We capture a tiny sliver of this flood. The rest bounces back into space, unused.

We are accustomed to thinking of energy as scarce and expensive. We fight wars over oil. We worry about carbon emissions. We debate whether we can afford to power our civilization without destroying the climate that makes civilization possible. Energy scarcity shapes our politics, our economics, our daily choices about what we can and cannot do.

But energy is not actually scarce. It only seems scarce because we live at the bottom of an atmosphere that filters and scatters sunlight, on a planet that rotates away from the sun for half of every day, under skies that cloud over unpredictably. By the time sunlight reaches a rooftop solar panel, it has lost most of its intensity. From the perspective of the solar system as a whole, energy is absurdly, incomprehensibly abundant. The sun radiates enough power to support trillions of human beings at high standards of comfort. We intercept less than a billionth of its output – and use only a fraction of what we intercept.

In free space, this abundance becomes directly accessible.

Solar collectors in orbit face the sun continuously – no night, no clouds, no atmosphere. The energy harvest per square meter is five to ten times what's possible on Earth's surface. A solar array the size of West Virginia, positioned in space, could meet all of humanity's current energy needs – continuously, forever. And there's no limit to how many collectors you can deploy. The materials to build them float freely in asteroids and on the lunar surface. The space to position them is effectively infinite. Once you've built the manufacturing capacity, you can keep building collectors until you have all the energy you could possibly want – and then keep building.

Energy is the foundation of everything else. With enough energy, you can do almost anything.

You can process raw materials into useful forms. You can manufacture anything that can be designed. You can recycle waste with perfect efficiency. You can maintain vast, complex ecosystems. You can support billions of people in comfort without strip-mining a single forest or poisoning a single river.

But energy alone isn't enough. You also need materials.

This is what I mean by "the abundance machine." It's not a single device. It's a system – a set of interlocking capabilities that, once established, can produce material wealth on a scale that makes scarcity obsolete.

Let me walk through how this works, step by step.

The first step is the hardest: establishing initial infrastructure in space. This means launching materials and equipment from Earth, at substantial expense, to build the first construction facilities, the first solar collectors, the first mining operations. This bootstrapping phase is relatively costly and slow. It's where we are now, more or less – taking tentative steps, launching satellites and space stations, developing the technologies that will eventually mature into something more.

The second step is where things start to change: using space resources to build more space infrastructure. The Moon's surface is rich in silicon, aluminum, iron, oxygen, and other useful elements. Near-Earth asteroids contain similar materials plus significant quantities of nickel, cobalt, and precious metals. Once you can mine and process these resources in space, you no longer need to launch everything from Earth. You can build new facilities using local materials, at a fraction of the investment.

The third step is where abundance begins: self-replicating production. Imagine a factory in space that can build copies of itself. You start with one factory. It builds another. Now you have two. They each build another. Now you have four. Then eight, sixteen, thirty-two. Within a relatively short time – years, not centuries – you have thousands of factories, all working in parallel, all powered by free sunlight, all processing freely available asteroid materials.

This isn't science fiction. Self-replicating systems are a well-studied concept in engineering. We haven't built them yet, but there's no physical law preventing them. The challenge is complexity – designing systems robust enough to copy themselves reliably in the harsh environment of space. It's a hard problem, but it's an engineering problem, not a physics problem. And we're getting better at solving such problems every year.

The fourth step is abundance itself: using this vast productive capacity to build whatever people need. Habitats. Farms. Forests. Cities. Art. Science. Leisure. Whatever human beings want, in quantities limited only by the imagination.

I want to pause here and address a reasonable skepticism.

"This sounds too good to be true," you might be thinking. "There must be a catch. There's always a catch."

And you're right to be cautious. Utopian promises have a poor track record. History is littered with schemes that were going to solve everything and ended up solving nothing – or making things worse.

But notice what I'm **not** claiming.

I'm not claiming that space abundance will solve all human problems. It won't make people wise, kind, or good. It won't resolve the deep questions of meaning and purpose that have always accompanied human existence. It won't prevent heartbreak, loss, or the ordinary suffering that comes with being alive. Material abundance is not the same as human flourishing, and I would never pretend otherwise.

I'm also not claiming that space abundance will happen automatically or easily. The bootstrapping phase is genuinely difficult. Arguably, it requires sustained investment, international cooperation, and decades of patient work. Many things could go wrong along the way. We could fail.

What I'm claiming is more limited but still profound: **material scarcity is not a permanent feature of the human condition**. It's a feature of our current situation – living on a single planet with finite resources and limited energy. That situation can change. The resources and energy exist, in quantities that dwarf anything on Earth. The question is whether we'll access them, and if so, how we'll use them.

The physics works. The engineering is plausible. The economics become favorable once the initial infrastructure is in place. The catch is not that abundance is impossible. The catch is that abundance doesn't distribute itself.

Think about what happened when Europeans encountered the Americas.

Here was a vast territory, rich in resources, largely unknown to the newcomers. From a certain perspective, it represented abundance – more land, more timber, more minerals, more opportunity than Europe could offer. And indeed, the wealth extracted from the Americas transformed the global economy. It funded empires, built cities, powered the Industrial Revolution.

But who benefited from this abundance?

Not the indigenous peoples, who were decimated by disease, dispossessed of their lands, and subjected to centuries of violence and exploitation. Not the enslaved Africans who were transported across the ocean to labor in mines and plantations. Not even most of the European settlers, many of whom lived hard lives of poverty and toil. The abundance of the Americas was captured by those with the power to capture it – colonial administrators, plantation owners, merchant companies, imperial governments. It made some people very rich. It left most people no better off than before, and many people far worse.

This is the pattern we must understand if we want space abundance to unfold differently.

Abundance does not automatically become shared abundance. Resources do not automatically benefit everyone just because they exist. The mere presence of wealth – even vast wealth, even effectively unlimited wealth – does not guarantee justice, equality, or flourishing for all. What matters is the system that determines who gets access to the wealth and on what terms.

On Earth, those systems developed over centuries, shaped by violence, conquest, and entrenched power. By the time anyone thought to ask whether the arrangements were fair, the arrangements were already locked in. The people who benefited from inequality had every incentive to preserve it, and they had the power to do so. Changing the system meant fighting against those with the most to lose.

Space offers something different. Not a guarantee of justice, but an opportunity for it. The systems haven't been built yet. The patterns haven't been set. We're at the beginning, not the middle or the end. If we're thoughtful now – if we design the foundational structures well – we might avoid locking in the same inequalities that have plagued every human civilization to date.

"But wait," you might say. "Won't the first people to get to space simply grab everything for themselves? Won't the rich and powerful establish the same hierarchies out there that they've established down here? What's to stop them?"

This is exactly the right question. And the honest answer is: nothing is certain to stop them. The outcome depends on choices, and choices can go either way.

But consider the factors working in favor of a different outcome.

First, space is not like land. On Earth, controlling territory means excluding others from a finite resource. There's only so much coastline, so much fertile soil, so much fresh water. If I have it, you don't. This zero-sum dynamic drives much of human conflict. In space, the resources are so vast that this dynamic breaks down. The asteroid belt contains enough material to build billions of GIF habitats. The sun provides enough energy to power them all. There's no need to fight over scraps when the feast is effectively unlimited.

Second, the technology of abundance tends toward accessibility. The most valuable resources in space aren't gold or diamonds – they're sunlight, common minerals, and the knowledge to use them. These aren't things that can be easily monopolized. Sunlight falls on everyone. Asteroids are scattered across vast regions. Technical knowledge, once developed, is hard to suppress. The inputs to abundance are intrinsically more democratic than the inputs to wealth on Earth.

Third, the initial space infrastructure is being built through international cooperation, not conquest. The Outer Space Treaty – which we'll discuss in detail later – established that space is the "province of all mankind," not subject to national appropriation. This principle has held for nearly sixty years. It's being tested by current developments, but it hasn't been abandoned. We have a foundation to build on.

Fourth, we have something previous generations lacked: foresight. We can see where things might go wrong. We can learn from history. We can make conscious choices about the kind of civilization we want to build, rather than letting patterns emerge haphazardly from unexamined assumptions and short-term interests. This book is part of that effort – trying to raise awareness, provoke thought, inspire action before the window closes.

None of this guarantees a good outcome. But it suggests that a good outcome is possible, if we work for it.

Let me paint two pictures of the future, say, fifty or a hundred years from now.

In the first picture, space development has proceeded without conscious design. The first companies and nations to establish space infrastructure have claimed the best orbital positions and the richest asteroids. They've built habitats, yes – but habitats they control, for people who can pay. The abundance of space has become another form of wealth, concentrated in the hands of those who got there first. Most humans still live on Earth, still struggling with scarcity, now looking up at a sky dotted with private paradises they'll never enter. The old inequalities have been projected onto a larger canvas. The opportunity has been squandered.

In the second picture, space development has been guided by deliberate principles. The infrastructure of abundance has been built as a common resource, accessible to all. GIF habitats have proliferated – thousands of them, then millions, each a self-governing community with its own character and culture. People move freely among them, seeking communities that fit their values and aspirations. Material needs are met for everyone; no one goes hungry, homeless, or without medical care. Human energy goes into art, science, exploration, relationships, meaning – the things that matter once survival is no longer a struggle. Earth is treasured and protected, no longer burdened with the impossible task of supporting all human ambitions. The abundance machine has become what it could always have been: a foundation for universal flourishing.

These two pictures represent different choices, not different technologies. The same engineering that enables the first picture also enables the second. The difference is in the principles we establish, the institutions we build, the decisions we make now while the patterns are still forming.

I've described the abundance machine in terms of energy and materials – sunlight and asteroids and self-replicating factories. But there's another component I haven't mentioned yet, one that makes everything else possible.

The abundance machine is not just solar collectors and asteroid mines. It's the entire orchestrated system – energy, materials, superautomation, organization – that makes universal flourishing possible for the first time in history.

I've been describing what's possible. But possibility is not destiny. The abundance machine won't build itself, and even if built, it won't automatically serve everyone.

The next chapter turns to this problem directly. If abundance doesn't share itself – if history shows that wealth tends to concentrate rather than spread – then what can we do about it? Is there any reason to think that space will be different from every previous frontier?

I believe there is. But it requires understanding not just the opportunity before us, but the dangers – and the choices we must make to navigate them.

☺

Part Two – The Fork in the Road

☺

Chapter 4: Abundance Doesn't Share Itself

In 1532, a small band of Spanish conquistadors encountered the Inca Empire.

The Incas controlled one of the largest and wealthiest civilizations on Earth. Their empire stretched well over two thousand miles along the western coast of South America. They had built roads, cities, temples, and agricultural systems of remarkable sophistication. They had storehouses filled with grain, textiles, and precious metals. By any measure, they possessed abundance.

Within a few years, that abundance belonged to Spain.

Francisco Pizarro captured the Inca emperor Atahualpa and held him for ransom. The ransom paid was staggering – a room filled with gold, twice over with silver. It remains one of the largest ransoms in human history. And after receiving it, Pizarro executed Atahualpa anyway, then proceeded to dismantle the empire piece by piece. The gold and silver of the Incas flowed across the Atlantic to fund Spanish wars, build Spanish palaces, and enrich Spanish nobles. The Inca people themselves were subjected to forced labor, disease, and systematic oppression that reduced their population by perhaps ninety percent within a century.

The abundance of the Incas did not protect them. It made them a target.

I tell this story not to relitigate colonial history but to illustrate a principle that runs through all of human experience: **abundance does not distribute itself**. The existence of wealth – even great wealth, even wealth beyond imagining – tells you nothing about who will benefit from it. That question is answered by power, and power follows its own logic.

This is the hardest lesson for optimists to accept. We want to believe that prosperity lifts all boats, that growing the pie helps everyone, that technological progress naturally bends toward justice. And sometimes, in some ways, it does. The average person today lives better than kings did a few centuries ago, at least in material terms. Progress is real.

But progress is not automatic, and it is not evenly shared. Every great expansion of human capability has been accompanied by new forms of exploitation, new concentrations of power, new ways for some to benefit at the expense of others. The Industrial Revolution created unprecedented wealth and also created dark satanic mills where children labored sixteen hours a day. The digital revolution connected the world and also enabled surveillance capitalism, algorithmic manipulation, and the hollowing out of entire industries. It seems progress and exploitation are not opposites, but are dance partners, moving together through history.

If we want space abundance to unfold differently, we need to understand why it so often unfolds the same way.

The pattern repeats because of a simple dynamic: those who control the means of production control the fruits of production.

This isn't a Marxist slogan. It's a descriptive observation about how economies work. Whoever owns the factories decides what they make and who gets the products. Whoever owns the land decides what grows there and who eats. Whoever owns the platforms decides what information flows and who profits from attention. Ownership is control, and control is power.

When new sources of abundance emerge, the crucial question is always: Who will own them?

Consider oil. In the early twentieth century, petroleum transformed from a curiosity into the lifeblood of industrial civilization. It powered cars, ships, airplanes, factories. It became the feedstock for plastics, fertilizers, pharmaceuticals. Whoever controlled oil controlled the future.

And who came to control oil? Not humanity in general. Not the people who lived above the oil deposits. A handful of companies – Standard Oil, Royal Dutch Shell, British Petroleum, and a few others – consolidated control over extraction, refining, and distribution. They became some of the most powerful economic entities in history. The nations where oil was found often remained poor, their wealth extracted and exported, their governments corrupted or overthrown when they tried to assert control over their own resources.

The abundance was real. The benefits were concentrated.

Or consider the internet. Here was a technology that promised to democratize information, connect humanity, give everyone a voice. And in some ways it delivered on that promise. But it also gave rise to a new class of oligarchs – the platform owners who sit at the chokepoints of digital life, extracting value from every interaction, accumulating wealth and power on a scale the robber barons of the Gilded Age could hardly imagine.

Again: the abundance was real. The benefits were concentrated.

Why does this keep happening? Why don't the benefits of abundance spread more evenly?

Several mechanisms are at work.

First movers capture the territory. When a new frontier opens – whether physical territory, a new technology, or a new market – those who arrive first have enormous advantages. They claim the best positions. They establish the initial rules. They accumulate resources that let them defend and extend their advantages. By the time others arrive, the game is already shaped in favor of the incumbents.

Network effects amplify concentration. Many modern systems exhibit increasing returns to scale. The more users a platform has, the more valuable it becomes, which attracts more users, which makes it more valuable still. The more factories you own, the cheaper you can produce goods, which lets you undercut competitors and buy more factories. These dynamics tend toward monopoly or oligopoly. The big get bigger; the rest get squeezed out.

Power defends itself. Once a group has accumulated wealth and control, they have strong incentives to preserve their position – and they have the resources to do so. They can influence laws, capture regulators, fund politicians, shape public opinion, and suppress alternatives. The playing field doesn't stay level because those ahead have every reason to tilt it further in their favor.

Ideology naturalizes inequality. Perhaps most insidiously, those who benefit from unequal arrangements tend to develop beliefs that justify those arrangements. The rich deserve their wealth because they're smarter, harder-working, more virtuous. The poor deserve their poverty because they're lazy, foolish, morally deficient. These beliefs make inequality seem natural and inevitable, discouraging challenges to the system.

None of this is conspiracy. It's just how systems evolve when no countervailing force prevents concentration. The default trajectory of abundance is toward capture by the few.

"But space is different," you might object. "You said so yourself. The resources are so vast that there's no need to fight over them. Sunlight falls on everyone. Asteroids are scattered everywhere. The zero-sum logic doesn't apply."

This is partly true, and it's the reason I have hope. But it's not automatically true. It's only true if we make it true.

Consider how space development is actually unfolding right now.

A handful of companies – SpaceX, Blue Origin, a few others – are developing the launch capabilities that will open access to space. They're doing remarkable work, driving down costs, achieving things that governments couldn't or wouldn't do. But they're private companies, owned by some of the wealthiest individuals on Earth, operating to generate returns for their shareholders.

When these companies start mining asteroids, who will own the minerals? When they build the first habitats, who will control access? When the infrastructure of space abundance starts to take shape, whose interests will it serve?

The United States passed a law in 2015 – the Commercial Space Launch Competitiveness Act – that explicitly grants American citizens and companies property rights over resources they extract from asteroids and other celestial bodies. Other nations are considering similar legislation. The legal framework for private ownership of space resources is being constructed right now, largely outside public awareness or debate.

I'm not saying private enterprise is bad. Private companies have driven much of the innovation that makes space development possible. Profit motives can align with broader benefits. Market mechanisms can allocate resources efficiently.

But private enterprise, left entirely to its own logic, will produce the same concentrations of wealth and power in space that it has produced on Earth. The first companies to establish asteroid mining operations will have enormous advantages over latecomers. Network effects will amplify those advantages. The owners of space infrastructure will become the new oligarchs, and the rest of humanity will relate to space abundance the way most humans have always related to abundance: as outsiders looking in, dependent on the decisions of those who control the means of production.

This is not inevitable. But it is the default. It's what will happen if we do nothing to prevent it.

There's another danger, perhaps even greater: military competition.

On Earth, the struggle for resources has always been intertwined with the struggle for military dominance. Nations arm themselves to protect what they have and to take what others have. The threat of violence lurks behind every negotiation, every treaty, every international arrangement. Peace, where it exists, is maintained by deterrence – the knowledge that aggression will be met with retaliation.

This system is unstable and brutal, but we've lived with it because we've seen no alternative. The conditions of planetary existence – finite territory, limited resources, inescapable proximity – seem to make competition inevitable.

Space could offer an escape from this trap. With unlimited resources and unlimited room, the zero-sum logic of military competition loses its foundation. There's no need to fight over territory when territory is effectively infinite. There's no need to secure resources by force when resources are abundant beyond any possible use.

But this peaceful potential will not realize itself automatically.

Right now, nations are extending their military competition into space. Anti-satellite weapons have been tested by multiple countries. Space is increasingly viewed as a domain of warfare, to be controlled and denied to adversaries. The same powers that have militarized every previous frontier are busy militarizing this one.

If this continues, the abundance of space will be overshadowed by the same dynamics of threat and counter-threat that have plagued Earth. Habitats will be vulnerable to attack. Resources will be contested by force. The vast potential of space will be channeled into the same destructive competitions that have consumed so much human energy throughout history.

We could build paradise, and instead build another battlefield.

I've painted a grim picture in this chapter. Intentionally so.

The optimistic case for space – that abundance will naturally produce flourishing – is seductive but naive. It ignores everything we know about how human systems actually develop. It assumes that space will be different from every previous frontier without explaining why or how.

The pessimistic case – that space will simply replicate Earth's patterns of exploitation and conflict – is arguably more realistic, but also not inevitable. It describes the default trajectory, not the only possible trajectory. Defaults can be changed. Patterns can be broken. History is not destiny.

The question is: What would it take to change the default?

This is where my story takes an unexpected turn. Because it turns out that we're not starting from scratch. We're not facing this challenge with nothing but good intentions and vague hopes. We have something concrete to work with – a foundation that was laid more than half a century ago, at the height of the Cold War, by people who understood that space was too important to leave to the default dynamics of competition and capture.

We have a treaty.

It's an imperfect treaty, incomplete in crucial ways, under pressure from developments its framers never anticipated. But it exists. It establishes principles. It provides a starting point.

In the next chapter, I'll tell you about this treaty – where it came from, what it says, and why it offers something like hope.

☺

Chapter 5: The Blank Canvas

Every previous frontier in human history came with baggage.

When Europeans arrived in the Americas, they encountered civilizations that had been developing for thousands of years. The land was not empty. It was home to hundreds of distinct peoples with their own languages, cultures, political systems, and claims to the territory. The "New World" was new only to the newcomers. For everyone already there, it was simply the world – the only world they had ever known.

The same was true everywhere humans expanded. Australia had been inhabited for sixty thousand years before the first European ships arrived. The Pacific Islands had been settled and resettled in successive waves of migration. Africa, Asia, Europe – every corner of every continent carried the weight of history, the accumulated claims and counterclaims of peoples who had lived there, fought there, built there, died there.

This meant that every frontier expansion involved collision. Newcomers encountered existing inhabitants. Different systems of ownership, governance, and meaning clashed. The results were often violent – conquest, displacement, genocide. Even when violence was avoided or minimized, the encounter between different ways of life created conflicts that took generations to resolve, if they were ever resolved at all.

Space is different.

There is no one there.

This simple fact changes everything. When we speak of settling space, we are not speaking of displacing anyone. There are no indigenous peoples of the asteroid belt. No ancient civilizations flourish on the moons of Jupiter. The Lagrange points where GIF habitats might orbit are empty – not temporarily empty, not empty because someone was driven out, but empty in the most absolute sense. No conscious being has ever called these places home.

This means something profound: for the first time in human history, we have the opportunity to build a new civilization without committing an original sin.

Every nation on Earth, without exception, was founded through some combination of conquest, displacement, and violence. The land we live on was taken from someone. The borders we defend were drawn through blood. The institutions we inherited carry the marks of old injustices, old compromises, old arrangements made by people who are long dead but whose choices still constrain us. We are all, in some sense, living in houses built on troubled foundations.

Space offers a blank canvas.

Not a blank canvas in the sense that anything is possible – physics constrains us, economics constrains us, human nature constrains us. But a blank canvas in the sense that we are not overwriting someone else's story. We are not tearing down to build up. We are not negotiating with the ghosts of prior inhabitants. We are starting, genuinely starting, with nothing but empty space and sunlight and rock.

The moral weight of this opportunity is staggering. For once, we can do it right from the beginning.

Consider what "doing it right from the beginning" might mean.

On Earth, we have inherited systems of property that evolved over centuries through violence and arbitrary seizure. Someone's ancestor took the land by force; their descendants inherited it; now we call it "ownership" and build legal systems to protect it. The original taking is laundered through time until it seems natural, inevitable, just the way things are.

In space, no one has taken anything yet. We can decide, collectively and deliberately, what property means in this new context. We can design systems that serve human flourishing rather than merely ratifying historical accidents of conquest.

On Earth, we have inherited borders that reflect ancient wars, colonial impositions, and diplomatic compromises made by people pursuing their own interests with no thought for those who would live with the consequences. These borders trap people in poverty, separate families, fuel conflicts that persist for generations.

In space, there are no borders yet. GIF habitats are separate structures; there's no contested boundary between them. We can design systems of movement and membership that respect human dignity and freedom rather than perpetuating arbitrary divisions.

On Earth, we have inherited political systems that evolved through struggle and compromise, carrying forward assumptions and structures from eras with very different challenges. Monarchies that became constitutional. Republics that retained aristocratic elements. Democracies distorted by wealth and manipulation. Every system is a palimpsest (new writing over old), never quite escaping the patterns laid down before.

In space, we can design governance from first principles. What do communities actually need? How should decisions be made? How can power be prevented from concentrating? How can individual freedom be protected while maintaining social cohesion? These questions can be asked freshly, without the constraint of having to reform existing institutions that resist change.

I want to be careful here not to sound naive.

Human beings will bring themselves to space, and human beings carry their own baggage – psychological, cultural, ideological. We don't become angels by changing our location. The settlers of every frontier have brought their prejudices, their bad habits, their capacity for cruelty and selfishness. Space settlers will be no different. They will squabble, compete, form factions, pursue narrow interests, make mistakes.

The blank canvas of space does not mean a blank slate for human nature.

But there's a difference between the baggage we carry in our minds and the baggage embedded in our institutions. Mental baggage is individual; it varies from person to person, and it can change through education, experience, and conscious effort. Institutional baggage is collective; it persists even when individuals want to change it, because institutions have their own momentum, their own defenders, their own resistance to reform.

Space offers us the chance to build new institutions without inherited baggage – institutions designed for abundance rather than scarcity, for cooperation rather than coercion, for freedom rather than control. The humans who inhabit these institutions will still be flawed, but they'll be flawed humans operating within structures designed to bring out their better qualities rather than their worse ones.

This is not utopia. It's just good design.

Think about the difference between renovating an old house and building a new one.

When you renovate, you're constrained by existing structures. The foundation determines what loads the walls can bear. The plumbing and wiring run through channels that are expensive to relocate. The layout reflects the needs and tastes of people who lived differently than you do. You can make improvements, but you're always working within limits set by previous builders.

When you build new, you start with your own needs. You can orient the house to capture sunlight optimally. You can design the layout for how you actually live. You can incorporate technologies and materials that didn't exist when the old house was built. The constraints are physics, budget, and imagination – not the accumulated decisions of previous generations.

Space is a new build. Earth is a renovation.

This doesn't mean the new build will automatically be better. Plenty of new houses are poorly designed, while some renovations produce wonderful results. The outcome depends on the skill, wisdom, and values of the builders. But the new build has degrees of freedom that the renovation lacks. It can be optimized in ways that the renovation cannot.

The question is whether we'll use those degrees of freedom wisely.

There's a concept in political philosophy called the "constitutional moment" – a rare period when the fundamental rules of a society are up for grabs, when choices can be made that will constrain all future choices.

Most of the time, we live within constitutional orders that we inherited. We can tinker at the margins, pass new laws, elect different leaders, but the basic framework is fixed. The constitution – written or unwritten – sets the boundaries of what's politically possible.

But occasionally, the framework itself becomes fluid. A revolution occurs. A war ends. A new nation is founded. In these moments, decisions that will bind generations are made by whoever happens to be in the room. The American founders, meeting in Philadelphia in 1787, made choices that still shape the lives of hundreds of millions of people. They could have chosen differently. Every path not taken represents a world that might have been.

Space is approaching its constitutional moment.

The fundamental rules haven't been written yet. The basic frameworks haven't been established. The patterns that will shape space civilization for centuries – perhaps for millennia – are still fluid, still open to influence.

This won't last. Once patterns are established, they become self-reinforcing. Once institutions exist, they develop constituencies that defend them. Once wealth and power accumulate in certain hands, those hands will fight to keep them. The window of possibility is open now, but it will close.

The blank canvas will be painted. The only question is what picture will emerge – and who will hold the brush.

I've been speaking abstractly about institutions and frameworks. Let me make it concrete.

Right now, in the mid-2020s, crucial decisions are being made about space governance. Some are being made explicitly, through laws and treaties and regulations. Others are being made implicitly, through the actions of companies and governments pursuing their own interests without much thought for long-term consequences – still others via policies shielded as top secret.

Who can own resources extracted from asteroids? Laws are being written right now that answer this question.

Who can control orbital positions – the prime real estate of near-Earth space? First-come-first-served practices are establishing precedents right now.

Who has authority to regulate activities in space? Jurisdictional frameworks are being contested right now.

What weapons can be deployed in orbit? The boundaries of acceptable military activity are being tested right now.

Each of these decisions, made today, will constrain what's possible tomorrow. Each precedent established now will be cited for generations. Each power grab that goes unchallenged will become the new normal.

The blank canvas is being marked. Tentatively so far, but increasingly boldly. We don't have unlimited time to decide what picture we want.

This brings us to a peculiar feature of our historical moment.

We are living through a transition that most people don't perceive. The decisions being made about space – in national legislatures, in international forums, in corporate boardrooms – will affect the lives of billions of people not yet born. But these decisions are being made with almost no public attention, no widespread debate, no sense of their true significance.

Ask the average person on the street about space governance, and you'll get a blank look. Ask them about the Outer Space Treaty, and they'll say they've never heard of it. Ask them who should own asteroid resources, and they'll wonder why it matters.

It matters because the blank canvas is the rarest of gifts – a chance to build something new without destroying something old. It matters because the window is closing, and once it closes, we'll live with the consequences indefinitely. It matters because this is our generation's unique responsibility, the thing we can do that no generation before could do and no generation after will be able to do again.

The settling of space will happen. The only question is whether it will happen thoughtfully or carelessly, with wisdom or without, for the benefit of all of humanity or for the benefit of whoever grabs first and fastest.

In the next chapter, I'll be more specific about the forces that are shaping space development right now – and why those forces, left unchecked, are likely to produce outcomes we don't want. The door of opportunity is still open, but it's beginning to swing shut.

Before we examine those threats, though, I want you to hold onto the vision of the blank canvas. Not as naive optimism, but as genuine possibility. The weight of history does not bind us here. The sins of the past need not be repeated. For this one moment, in this one domain, we have the chance to choose consciously what kind of civilization we want to build.

That chance is precious. We must not waste it.

☺

Chapter 6: The Door Is Closing

On November 15, 2021, Russia destroyed one of its own satellites.

The target was Cosmos 1408, a defunct Soviet-era surveillance satellite that had been orbiting Earth since 1982. A ground-launched missile streaked up from the Plesetsk Cosmodrome, crossed the boundary of space, and struck the satellite at a combined velocity of several kilometers per second. Cosmos 1408 ceased to exist as a coherent object. In its place, a cloud of debris – at least 1,500 fragments large enough to track, and countless smaller pieces – began spreading through low Earth orbit.

The International Space Station, with seven people aboard, passed near the debris field multiple times in the following days. Astronauts and cosmonauts sheltered in their docked spacecraft during the closest passes, ready to evacuate if debris struck. Indeed, the debris will remain in orbit for years, perhaps decades, threatening every satellite and crewed mission that passes through the affected region.

Russia was not the first country to demonstrate this capability. China destroyed a satellite in 2007, creating an even larger debris field that still clutters orbit today. The United States did it in 2008. India in 2019. Each test added to the growing population of space junk – a population that threatens the very infrastructure that makes modern life possible.

This is not abstract. This is happening now.

The weaponization of space has been underway for decades, but it's accelerating.

During the Cold War, both superpowers developed anti-satellite weapons, tested them, and eventually stepped back from the brink. The Outer Space Treaty of 1967 banned nuclear weapons in orbit, and a kind of tacit restraint limited the militarization of space even as both sides used satellites for reconnaissance, communication, and early warning.

That restraint is eroding.

The United States established a Space Force in 2019 – a new branch of the military dedicated to space operations. China and Russia have reorganized their own military space capabilities. All three powers are developing not just anti-satellite missiles but a range of other weapons: ground-based lasers that can blind or damage satellites, maneuverable "inspector" satellites that can approach and interfere with other spacecraft, cyber capabilities that can disrupt satellite control systems.

The logic driving this arms race is familiar from every previous military competition. Each side sees the other's capabilities as threats that must be countered. Each defensive measure looks offensive from the other side. Each escalation justifies further escalation. The result is a spiral that no one wants but no one knows how to stop.

Space is becoming a warfighting domain – a place where future conflicts will be won or lost. Military planners speak openly about "space control" and "space denial" (tactics aimed to dominate the "high ground"). The satellites that guide missiles, coordinate troops, enable communications, and gather intelligence are now targets. The first hours of any major war between great powers would likely see attacks on space assets.

This changes everything.

Consider what a militarized space means for the vision of abundance I described earlier.

GIF habitats are large structures – miles across, home to thousands or millions of people. They're built for permanence, for comfort, for human flourishing. They are not built for war.

In a militarized space environment, every habitat is a target. Every community floating in the void is vulnerable to attack by adversaries who might be millions of miles away or just over the horizon. The same technologies that could destroy a satellite could destroy a habitat – or hold it hostage.

How do you build a civilization under these conditions?

You don't build open, trusting communities. You build fortresses. You invest in defenses rather than gardens. You design for survival rather than flourishing. You organize your society around the threat of violence rather than the pursuit of happiness.

This is the future we're drifting toward. Not because anyone wants it, but because no one is stopping it. The military competition that has consumed so much human energy on Earth is extending into the one domain where we had a chance to do something different.

The military threat is the most dramatic, but it's not the only way the door is closing.

The commercial scramble for space is intensifying just as rapidly. And while commerce is generally less destructive than warfare, unregulated commercial competition can also lock in patterns we'll regret.

Consider the race for orbital positions.

The most valuable real estate in near-Earth space is limited. Geostationary orbit – the ring around Earth where satellites can hover over a fixed point on the surface – has only so many slots. The prime locations for space stations, fuel depots, and manufacturing facilities are contested. The first to occupy a position has a significant advantage over latecomers.

Right now, companies and nations are rushing to claim these positions. SpaceX alone has launched thousands of Starlink satellites, filling low Earth orbit with a constellation that provides internet service – and establishes a presence that competitors must work around. Other companies are planning their own mega-constellations. The sky is being carved up, and the carving is happening faster than any regulatory framework can manage.

Or consider asteroid mining.

The first company to successfully extract resources from an asteroid will have demonstrated a capability of immense value. They'll have established techniques, equipment, and expertise that others will need years to replicate. They'll have a head start on the path to space-based manufacturing. And under current legal frameworks – particularly the U.S. Commercial Space Launch Competitiveness Act of 2015 – they'll own what they extract.

This isn't necessarily bad. Private enterprise has driven much of the innovation that makes space development possible. Profit motives can align with broader social benefits. The problem isn't commerce itself but unregulated commerce – commerce without guardrails, without mechanisms to ensure that the benefits are widely shared.

The pattern from Earth is familiar. When a new domain opens up – whether the internet, the genome, or artificial intelligence – the early movers establish positions that become increasingly unassailable. Regulatory frameworks lag years or decades behind technological capabilities. By the time society wakes up to the need for rules, the game is already over. The winners write the rules to protect their winnings.

Space is following the same pattern. The blank canvas is being painted, and most of humanity isn't even aware that brushes are moving.

There's a deeper problem still: the erosion of the international framework that was supposed to govern space.

The Outer Space Treaty of 1967 established foundational principles – that space should be used for peaceful purposes, that it belongs to all of humanity, that celestial bodies cannot be claimed by nations. These principles have held for nearly sixty years, longer than many people expected.

But the treaty is showing its age.

It was written in an era when only two nations had significant space capabilities, when the technologies we now take for granted were barely imaginable, when the prospect of commercial space development seemed remote. The treaty addresses the concerns of 1967, not the concerns of today.

Does the prohibition on national appropriation of celestial bodies prevent private companies from owning extracted resources? The treaty doesn't say clearly. Different nations are interpreting it differently, and the lack of consensus is creating legal uncertainty that some exploit while others worry.

Does the requirement for "peaceful purposes" prohibit military activities short of actual weapons? Again, ambiguity. Nations are pushing boundaries, testing limits, finding loopholes.

What enforcement mechanisms exist if a nation violates the treaty? Essentially none. The treaty relies on good faith and mutual interest. When those erode, the treaty's constraints erode with them.

I'll discuss the Outer Space Treaty in detail in the next chapter – its origins, its content, its remarkable success against long odds. For now, the point is simpler: the international framework for space governance is under pressure at exactly the moment when it most needs to be strengthened.

Let me be concrete about the timeline.

The next ten to twenty years are decisive.

During this period, the major space powers will make choices that shape everything that follows. Will they extend the weapons ban to include conventional weapons, or will they continue the arms race into orbit? Will they establish fair frameworks for resource extraction, or will they allow first-movers to capture everything? Will they strengthen international cooperation, or will they retreat into ultracompetition and conflict?

These choices are being made right now, often without public awareness or debate. By the time the general public understands what's at stake, the patterns may already be locked in.

I think of it like climate change, but compressed.

With climate change, we had decades to see the problem developing. Scientists warned us in the 1960s and 1970s. We had time – not unlimited time, but time – to change course. We didn't use that time well, and now we face consequences that could have been avoided.

With space governance, the timeline is shorter. The decisions being made today will have consequences within years, not decades. And unlike climate change, where the physical processes are slow-moving, space development is driven by human choices that can move very fast. A single nation deciding to withdraw from treaties, a single company achieving a technological breakthrough, a single military confrontation that escalates – any of these could shift the trajectory abruptly.

We don't have the luxury of waiting to see how things develop. By the time the trajectory is clear, it may be too late to change it.

I'm aware that this chapter has been heavy with warnings and alarm. I want to explain why I think the alarm is warranted.

Throughout this book, I've tried to present a vision of what's possible – a civilization of abundance, peace, and freedom, built on the blank canvas of space. This vision is not fantasy. It's grounded in physics, engineering, and a realistic assessment of human capabilities. The resources exist. The technologies are developing. The opportunity is real.

But opportunity is not destiny. The same physics that enables GIF habitats also enables weapons that could destroy them. The same technologies that could produce abundance could be captured by the few. The same blank canvas that allows for new beginnings could be painted with the same old patterns of domination and strife.

The hopeful version of outer space I presented earlier in this book is one possible future. But it's not the default future. The default is something darker – not because anyone chooses darkness, but because light requires active effort while darkness just requires yielding to the gravity of our worst habits or instincts.

I'm trying to create urgency without despair. The door is closing, but it's not closed. The patterns are forming, but they're not yet fixed. We have agency – real agency – to influence how this unfolds. But that agency must be exercised soon, before the window shuts.

So what can be done?

The problems I've described – military competition, commercial capture, erosion of international frameworks – are large and systemic. They involve great powers, vast economic interests, and entrenched institutional dynamics. No individual can solve them alone.

But individuals acting together, focused on the right leverage points, can shift trajectories.

The leverage point I want to focus on is the Outer Space Treaty itself – not as a fixed document but as a foundation that can be built upon. The treaty established principles that point toward the future we want. Those principles need to be strengthened, extended, and given teeth.

In the next chapter, I'll tell the story of how this treaty came to exist – how, at the height of the Cold War, with the world's two superpowers competing fiercely for advantage, humanity managed to agree that space should be different. It's a remarkable story, and it offers lessons for what we might accomplish now.

The door is closing. But it's not closed yet.

☺

Part Three – The Treaty We Already Have

☺

Chapter 7: A Miracle from 1967

On October 4, 1957, the Soviet Union launched a small metal sphere into orbit around the Earth. I was ten years old. In wonder, we all pointed toward the night sky, watching a new "star" glide across the firmament.

Sputnik weighed less than two hundred pounds. It carried no weapons, no cameras, no scientific instruments of consequence. All it did was circle the planet every ninety-six minutes, emitting a simple radio beep that amateur operators around the world could pick up on their receivers.

That beep terrified the West.

If the Soviets could put a satellite in orbit, they could put a nuclear warhead there. If they could reach space, they could rain destruction down from above with impunity. The same rockets that launched Sputnik could launch missiles that would cross continents in minutes, unstoppable by any defense. The strategic balance that had kept an uneasy peace since World War II suddenly seemed precarious.

The United States responded with a crash program to catch up. NASA was founded. Budgets exploded. The Space Race began in earnest – not as a scientific endeavor or a quest for knowledge, but as a military and ideological competition between two superpowers who genuinely believed the other might destroy them.

This was the context in which the 1967 Outer Space Treaty was negotiated. Not a time of peace and cooperation, but a time of fear, suspicion, and barely contained hostility. The Cuban Missile Crisis of 1962 had brought the world to the brink of nuclear annihilation. Proxy wars raged across Asia, Africa, and Latin America. The Iron Curtain divided Europe. Each side was convinced the other was bent on world domination.

And yet, somehow, in this environment of maximum distrust, the United States and the Soviet Union – along with dozens of other nations – agreed on a treaty to keep space peaceful.

How did that happen?

The answer begins with fear.

Both superpowers recognized that space offered terrifying new possibilities for warfare. Orbital nuclear weapons could strike with almost no warning. Anti-satellite weapons could blind the other side, leaving them unable to detect an incoming attack. Space-based platforms could provide decisive military advantages to whoever controlled them.

An arms race in space seemed inevitable. And both sides understood, at some level, that such a race would be catastrophically destabilizing. The nuclear balance on Earth was already precarious enough. Extending that competition into a new domain – one with fewer rules, less predictability, and faster timelines – could push both sides toward the hair-trigger postures that make accidental war almost inevitable.

There was also the matter of cost. Both nations were already spending enormous sums on nuclear weapons, conventional forces, and the early space programs. A full-scale militarization of space would require investments that even superpowers might struggle to sustain. If both sides could agree to limit the competition, both would benefit.

Fear and exhaustion are not noble motives. But they are powerful ones. And in the early 1960s, they created an opening for diplomacy.

A key figure on the American side was President Lyndon Johnson.

Johnson had been deeply involved in space policy since the Sputnik shock. As Senate Majority Leader in the late 1950s, he had held hearings, pushed for increased funding, and positioned himself as a leader on the issue. He understood both the military implications of space and its potential for peaceful cooperation.

In 1966, President Johnson repeatedly framed the choice starkly. The nations of the world, he stated, faced a decision: they could extend their conflicts into space, creating a new arena for the arms race that threatened humanity's survival. Or they could choose a different path – agreeing that space would be preserved for peaceful purposes, for the benefit of all.

The Soviets, despite their ideological hostility to the West, were receptive. Premier Alexei Kosygin had his own reasons for wanting to limit the space competition. The Soviet economy was straining under military burdens. A treaty that constrained both sides equally would allow resources to be redirected elsewhere.

Negotiations began at the United Nations, with both superpowers submitting draft treaties. The process was contentious – there were real disagreements about what should be prohibited and what should be permitted. But the underlying interest in reaching agreement was strong enough to overcome the obstacles.

On January 27, 1967, the Treaty on Principles Governing the Activities of States in the Exploration and Use of Outer Space, Including the Moon and Other Celestial Bodies – known ever after as the Outer Space Treaty – was opened for signature in Washington, London, and Moscow simultaneously.

The United States, the Soviet Union, and the United Kingdom – as depositary governments – led the signing, and representatives of dozens of other nations signed that same day. Today, over 115 countries are parties to the treaty, including every spacefaring nation.

On January 27, 1967, at the signing of the Outer Space Treaty, President Johnson's live remarks were remarkable: "We are taking the first firm step toward keeping outer space free forever from the implements of war. ... We have never succeeded in freeing our planet from the implements of war. But if we cannot yet achieve this goal here on earth, we can at least keep the virus from spreading."

So, what does the treaty actually say?

The core provisions are surprisingly clear and sweeping.

Space is free for exploration and use by all nations. No country can claim sovereignty over space, the Moon, or any other celestial body. The old patterns of territorial conquest – planting flags and claiming land – do not apply. Space is, in the treaty's phrase, "the province of all mankind."

Weapons of mass destruction are prohibited in space. Nations cannot place nuclear weapons or other WMDs in orbit, on celestial bodies, or anywhere else in outer space. This was the provision driven most directly by Cold War fears – the nightmare of orbital nuclear platforms that could strike without warning.

Celestial bodies must be used exclusively for peaceful purposes. The Moon and other celestial bodies cannot host military bases, weapons testing, or military maneuvers. The treaty creates, in effect, a demilitarized zone encompassing every solid body in the solar system.

Astronauts are envoys of mankind. Whatever their nationality, people in space are to be regarded as representatives of all of humanity. Nations must render them assistance in case of emergency, regardless of political relationships.

Nations are responsible for their space activities. Whether conducted by government agencies or private companies, a nation's space activities are the responsibility of that nation. This provision anticipated, in a limited way, the rise of commercial space development.

Nations must avoid harmful contamination. Space activities should not contaminate celestial bodies or Earth's environment. This provision has become increasingly relevant as we consider planetary protection and the disposal of space debris.

The treaty is not long – only seventeen articles. It establishes principles rather than detailed regulations. It creates no enforcement mechanism, no international space police, no penalties for violation. It relies, ultimately, on the good faith of its signatories and the mutual benefits of compliance.

And yet it has held.

That last point deserves emphasis. The Outer Space Treaty has held.

For nearly sixty years, through the remainder of the Cold War, through the collapse of the Soviet Union, through the rise of new space powers like China and India, through the emergence of commercial space companies, through transformations in technology that the treaty's framers could never have imagined – through all of this, the core provisions of the treaty have been respected.

No nation has placed nuclear weapons in orbit. No nation has claimed sovereignty over the Moon or any other celestial body. The principle that space is the province of all mankind remains, at least officially, the foundation of international space law.

This is remarkable.

Think about how many international agreements have failed, how many treaties have been violated or abandoned, how many noble principles have been discarded when they became inconvenient. The twentieth century is littered with broken promises and shattered frameworks. And yet this one treaty, negotiated at the height of superpower hostility, has endured.

Why?

Partly because the mutual interest in avoiding space weapons remained strong throughout the Cold War. Neither side wanted to trigger an arms race it might lose.

Partly because the treaty's provisions were clear enough to guide behavior but flexible enough to accommodate changing circumstances. The treaty didn't try to regulate every detail; it established principles that could be interpreted and applied as situations evolved.

Partly because the treaty created a shared expectation – a sense that space was different, that the normal rules of competition didn't fully apply there, that humanity had collectively decided to try something new. This shared expectation became self-reinforcing. Each year that passed without violation made the norm stronger.

And partly, perhaps, because of something harder to quantify: a genuine aspiration, even among hardened Cold Warriors, that space might represent a better future. The astronauts and cosmonauts who ventured into orbit reported transformative experiences – seeing Earth as a single, fragile whole, without the borders that divided it politically. Some of that vision filtered back into the diplomatic process, lending the treaty a moral weight beyond its strategic rationale.

I don't want to romanticize what happened. The treaty was not an act of idealism triumphing over realism. It was a calculated agreement among nations pursuing their own interests. Both superpowers retained the ability to militarize space if they chose; they simply agreed not to – for now.

Nor did the treaty prevent all military activity in space. Reconnaissance satellites, communications satellites, early warning satellites – all of these military uses continued and expanded. The treaty banned weapons of mass destruction and military activities on celestial bodies, but it left a great deal of gray area that both sides exploited.

And the treaty had significant gaps from the beginning. It didn't address conventional weapons in orbit. It didn't establish verification mechanisms. It didn't anticipate the commercial space industry or the legal questions that would arise around resource extraction. It was, and is, an incomplete framework.

But incomplete is not the same as worthless. The Outer Space Treaty established a foundation. It articulated principles. It created a starting point for everything that followed. Without it, the legal and diplomatic landscape of space would be far more chaotic than it is today.

The treaty proved that international cooperation on space is possible – even in the worst of circumstances, even between mortal enemies, even when the stakes are existential. If the United States and the Soviet Union could agree on the Outer Space Treaty in 1967, then surely we can build on that foundation today.

There's a lesson here about how change happens.

Grand visions are important. The dream of a peaceful, abundant space civilization – the vision I've been developing throughout this book – matters. It gives us something to aim for, a picture of what's possible, a reason to care about decisions that might otherwise seem abstract and distant.

But grand visions don't implement themselves. They become real through specific, concrete steps – treaties, laws, institutions, agreements. The Outer Space Treaty is not the full realization of any grand vision. It's a single step, taken at a specific moment, under specific circumstances, by specific people who saw an opportunity and seized it.

We need more steps like that. The treaty established that space should be peaceful and should belong to all of humanity. Now we need to extend those principles, strengthen them, give them teeth. The blank canvas I described earlier won't stay blank. The question is whether we'll fill it deliberately, guided by principles we've chosen, or haphazardly, driven by unrestrained competition and short-term interests.

The Outer Space Treaty shows that deliberate action is possible. It shows that nations can agree on principles even when they agree on little else. It shows that frameworks established at the right moment can endure for generations.

We are at another such moment. The decisions made in the next decade will shape space civilization for centuries. The Outer Space Treaty gave us a foundation. What we build on that foundation is up to us.

In the next chapter, I'll examine more closely what the treaty got right – the principles that point toward the future we want. Then we'll look at what the treaty needs now: the extensions and strengthening that could turn its promise into reality.

The miracle of 1967 was real. But miracles need tending. The door that was opened then could still close. It's our job to keep it open – and to walk through it.

☺

Chapter 8: What the Treaty Got Right

The Outer Space Treaty is not a long document. You can read it in fifteen minutes. Its language is general, its provisions broad, its enforcement mechanisms nonexistent. Compared to the detailed regulatory frameworks that govern other domains – international trade, aviation, the law of the sea – it seems almost naive in its simplicity.

And yet that simplicity may be its greatest strength.

The treaty's framers understood something that we often forget: foundational documents work best when they establish principles rather than regulations. The U.S. Constitution doesn't tell Congress exactly what laws to pass; it creates a framework within which legislation happens. The Universal Declaration of Human Rights doesn't specify every protection that every nation must provide; it articulates ideals meant to guide specific implementations.

The Outer Space Treaty works the same way. It doesn't try to anticipate every situation that might arise in space development – an impossible task in 1967 and probably impossible today. Instead, it establishes principles that can guide decision-making as circumstances evolve.

Those principles, properly understood, point directly toward the future we want.

Principle One: Space is the province of all mankind.

This phrase appears in Article I of the treaty, and it's worth pausing over. The exploration and use of outer space "shall be carried out for the benefit and in the interests of all countries, irrespective of their degree of economic or scientific development, and shall be the province of all mankind."

Notice what this does and doesn't say.

It doesn't say that space belongs to no one. It says space belongs to everyone. This is a positive claim, not merely a negative one. Space is not a vacuum of ownership waiting to be filled by whoever gets there first. It is already owned – by humanity as a whole.

This principle directly contradicts the logic that governed every previous frontier. When Europeans encountered the Americas, they claimed ownership based on discovery and conquest. When nations divided Africa at the Berlin Conference of 1884, they asserted sovereignty based on occupation and control. The pattern was always the same: empty spaces (declared empty by ignoring existing inhabitants) were carved up by the powerful.

The Outer Space Treaty says: not this time.

Space cannot be carved up. It cannot be claimed by nations or parceled out to favored interests. It belongs, already and permanently, to all of humanity – including future generations not yet born, including nations that don't yet exist, including people who will never travel there themselves.

This is a revolutionary principle. We haven't fully absorbed its implications.

If space is the province of all mankind, then the resources of space – the sunlight, the asteroids, the orbital positions – are common heritage. They don't belong to whoever extracts them first. They belong to everyone. The benefits of space development should flow to all, not just to those with the wealth and technology to get there.

Now, the treaty doesn't spell out exactly what this means in practice. That's one of its gaps, and I'll discuss it in the next chapter. But the principle is clear. Whatever specific arrangements we develop for space resources, they must be consistent with the idea that space belongs to all of us.

Principle Two: Space shall be used for peaceful purposes.

The treaty's preamble speaks of "the common interest of all mankind in the progress of the exploration and use of outer space for peaceful purposes." Article IV prohibits weapons of mass destruction in orbit and requires that celestial bodies be used "exclusively for peaceful purposes."

This principle has been tested and contested over the decades. What counts as "peaceful"? Military reconnaissance satellites are permitted – they're considered stabilizing because they reduce uncertainty. But what about anti-satellite weapons? What about missile defense systems based in space? The boundaries are fuzzy, and different nations interpret them differently.

But the core principle remains powerful: space should not become a battlefield.

This principle matters even more today than it did in 1967. Back then, space was primarily a domain of government activity – national space agencies conducting exploration and military services operating satellites. The potential for space-based civilization was theoretical.

Now it's becoming practical. We're approaching the era when significant numbers of people might actually live permanently in space, when space infrastructure might become essential to human civilization, when the stakes of space conflict would be measured not just in destroyed hardware but in destroyed communities.

In this context, the principle that space should be peaceful becomes existential. You cannot build a thriving civilization under constant threat of attack. You cannot invest in long-term infrastructure if it might be destroyed in the next conflict. You cannot raise children and plan for the future in a war zone.

The peaceful purposes principle is not just a nice aspiration. It's a precondition for everything else we want to achieve in space. If space becomes militarized – if the weapons competition that has consumed so much human energy on Earth extends into orbit – then the vision of abundant space civilization becomes impossible.

The treaty got this right. It recognized, even in 1967, that peace is not optional. It's foundational.

Principle Three: No national appropriation of space or celestial bodies.

Article II of the treaty states: "Outer space, including the Moon and other celestial bodies, is not subject to national appropriation by claim of sovereignty, by means of use or occupation, or by any other means."

This is one of the clearest provisions in the treaty, and one of the most consequential.

Throughout human history, territorial expansion has followed a predictable pattern. A powerful nation sends explorers or settlers into a new region. They plant a flag, build a fort, establish a presence. The nation then claims sovereignty over the territory – the exclusive right to govern, to make laws, to control who enters and what happens there. Other nations either accept this claim, contest it through diplomacy, or fight wars over it.

The Outer Space Treaty breaks this pattern completely. No nation can claim sovereignty over any part of space. Not the Moon, not Mars, not the asteroids, not orbital regions around Earth. It doesn't matter who gets there first. It doesn't matter who plants a flag or builds a base. Sovereignty – the foundation of territorial control on Earth – simply doesn't apply.

Think about what this means.

On Earth, every square inch of land (with few exceptions) belongs to some nation. Where you are determines whose laws apply to you, whose authority governs you, whose military can compel you. Borders matter immensely. Wars are fought over them. Lives are constrained by them.

In space, under the treaty's framework, there are no borders. There is no territory to conquer, no sovereignty to extend, no imperial domain to establish. Nations retain jurisdiction over their own spacecraft and citizens, but they cannot claim the space itself.

This principle creates the possibility of something genuinely new: communities in space that are not extensions of terrestrial empires, not colonies of existing nations, not subject to the same patterns of domination that have shaped every society on Earth.

The principle is not self-enforcing. There are nations and corporations that would like to establish de facto control over valuable regions of space, whether or not they use the language of sovereignty. The principle needs to be defended and extended. But the principle exists. It's been agreed to. It provides a foundation.

Principle Four: Astronauts as envoys of mankind.

Article V describes astronauts as "envoys of mankind in outer space." It requires nations to render them all possible assistance in case of emergency, to return them safely if they land in foreign territory, and to inform other nations of any phenomena in space that could endanger human life.

This provision might seem like a practical measure for emergencies, and it is that. But it's also something more.

By declaring astronauts "envoys of mankind," the treaty asserts that people in space represent all of humanity, not just their own nations. A Chinese taikonaut, an American astronaut, a Russian cosmonaut – whatever their flags and loyalties on Earth, in space they are envoys of our entire species.

This principle has been honored in practice. When emergencies have occurred in space, international cooperation has kicked in regardless of political tensions on the ground. The collaborative spirit aboard the International Space Station – where Americans, Russians, Europeans, Japanese, and Canadians have worked together even when their governments were at odds – reflects this principle.

But the principle points toward something larger. If astronauts are envoys of mankind, then space communities should perhaps be understood similarly – not as colonies of particular nations but as experiments in human civilization carried out on behalf of all of humanity.

This is a seed that hasn't fully sprouted. The treaty doesn't develop the idea or spell out its implications. But the seed is there, planted in 1967, waiting for conditions that allow it to grow.

Principle Five: International responsibility and cooperation.

Several articles of the treaty address how nations relate to each other in space. Nations are responsible for their national activities in space, whether conducted by government agencies or private entities. Nations must conduct space activities with due regard for the interests of other nations. Nations must consult with each other if their activities might cause harmful interference. Nations must allow other nations to observe their space activities.

These provisions establish that space is a shared domain requiring cooperation, not a lawless frontier where anything goes.

On Earth, we're accustomed to thinking of international relations as fundamentally competitive – nations pursuing their own interests, cooperating only when it serves them, always ready to defect when advantage beckons. This model has some truth to it. Nations do compete. They do pursue their own interests.

But the Outer Space Treaty embeds a different model into the foundations of space law. It assumes that space activities affect everyone, that nations must take each other into account, that cooperation is not optional but required. The treaty doesn't just permit international cooperation in space; it mandates it.

This principle becomes more important as space activities intensify. When only a few nations had any space capability, coordination was relatively simple. Now dozens of countries operate satellites. Private companies are launching thousands of spacecraft. The potential for interference, collision, and conflict grows with every passing year.

The treaty's framework of international responsibility and cooperation provides the foundation for managing these challenges. It's not sufficient by itself – more detailed agreements and institutions are needed – but it establishes that space is a domain requiring collective governance, not unilateral action.

Principle Six: Freedom of exploration and scientific investigation.

Article I declares that "outer space, including the Moon and other celestial bodies, shall be free for exploration and use by all States without discrimination of any kind." Article II guarantees "freedom of scientific investigation in outer space."

These provisions establish that space is open. No nation can close it off, restrict access to it, or prevent others from conducting legitimate activities there. The default is freedom – freedom to explore, to investigate, to use – limited only by the other principles the treaty establishes.

This is the positive complement to the prohibition on national appropriation. Space cannot be claimed because it's already everyone's. And because it's everyone's, everyone has the right to use it. Small nations and large, wealthy nations and poor, established powers and newcomers – all have equal rights to access and use outer space.

In practice, of course, capability matters. A nation without rockets cannot exercise its right to access space. But the principle is important nonetheless. It means that as capabilities spread – as more nations develop space programs, as private companies lower the cost of access – the benefits of space can spread too. No one has the right to exclude others from the common heritage of humanity.

Taken together, these principles outline a vision of space that differs radically from the patterns of terrestrial history.

Space belongs to all of humanity. It must be used peacefully. It cannot be carved up into national territories. The people who venture there represent our entire species. Nations must cooperate and take responsibility for their actions. Everyone has the right to access and explore.

This is not a complete framework for space civilization. It leaves many questions unanswered – questions about property rights, about governance structures, about enforcement mechanisms, about the specific rules that should govern specific activities. The treaty's framers knew they couldn't answer all these questions in 1967. They established principles and left the details for future generations.

We are those future generations. The principles have been established. Now we must build on them.

In the next chapter, I'll turn to what the treaty lacks – the gaps and ambiguities that need to be addressed if the promise of the principles is to be fulfilled. The foundation is solid. But a foundation is only the beginning.

☺

Chapter 9: What the Treaty Needs Now

The Outer Space Treaty is like a house with a magnificent foundation but missing walls.

The foundation is solid. The principles I described in the last chapter – space as the province of all mankind, peaceful purposes, no national appropriation, international cooperation – these are, in spirit, the right principles. They point toward the future we want. They've held up remarkably well for nearly sixty years.

But principles alone don't build a civilization. You need structures, rules, mechanisms, institutions. You need the walls, the roof, the plumbing, the wiring. You need the practical arrangements that turn grand ideals into lived reality.

The Outer Space Treaty didn't provide these. It couldn't, really – not in 1967, when space development was in its infancy and the specific challenges of the future were impossible to foresee. The treaty's framers did what they could: they established the foundation and trusted future generations to build the rest.

We are those future generations. The building has begun, whether we're ready or not. The question is whether we'll construct something worthy of the foundation – or let the edifice rise haphazardly, shaped by whoever happens to be holding a hammer.

Gap One: Conventional weapons are not prohibited.

This is the most dangerous omission in the treaty.

Article IV bans weapons of mass destruction from space. No nuclear weapons in orbit. No chemical or biological weapons on celestial bodies. This prohibition has held, and it matters enormously. A nuclear arms race in space would have been catastrophic.

But conventional weapons – missiles, lasers, kinetic impactors, electronic warfare systems – are not clearly prohibited. The treaty requires that celestial bodies be used "exclusively for peaceful purposes," but it's ambiguous about orbital space itself. Nations have interpreted this ambiguity to permit military activities short of WMDs.

The result is the creeping militarization I described in Chapter 6. Anti-satellite weapons have been tested. Space forces have been established. Military planners openly discuss "space control" and "space denial" (tactics aimed to dominate the "high ground"). The taboo against treating space as a warfighting domain is eroding.

This erosion threatens everything.

If conventional weapons proliferate in space, the peaceful civilization we envision becomes impossible. Every habitat becomes a target. Every orbital position becomes contested territory. Every investment in long-term infrastructure becomes a gamble that war won't destroy it.

In principle, the solution is straightforward: extend the weapons ban to include all weapons, not just weapons of mass destruction. Make space a weapons-free zone, period. (No weapons, no weapons-making, no weapons research.)

This would not prevent nations from proactively using space for war-prevention functions – reconnaissance, communication, navigation. These activities are stabilizing; they reduce uncertainty and help prevent wars.

A satellite that takes photographs is not a weapon. A satellite meant to collide with a competitor's satellite is a weapon. A ground station that communicates with spacecraft is not a weapon. A ground-based laser designed to blind satellites is a weapon. The line can be drawn.

Drawing it won't be easy. There are verification challenges. There are definitional challenges – some technologies have both peaceful and military uses. There are political challenges – the various vested interests.

But difficult is not impossible. Arms control agreements have navigated similar challenges before. The key is political will – a shared recognition that everyone benefits from keeping space weapons-free, even nations that might gain temporary advantage from deploying them.

The Outer Space Treaty established that space should be peaceful. Now we need to make that principle operational by banning the weapons that would make it violent.

Gap Two: Resource rights are undefined.

The treaty says that space cannot be subject to national appropriation. But what about private appropriation? What about resources that are extracted from celestial bodies and removed?

The treaty is silent on these questions. In 1967, asteroid mining and lunar resource extraction were science fiction. No one thought to address them.

Now they're approaching reality. Companies are developing plans to mine asteroids for water, metals, and other valuable materials. Nations are eyeing the Moon's resources – particularly water ice at the poles, which could be converted to rocket fuel. The question of who owns what is no longer theoretical.

Into this vacuum, nations have begun asserting their own answers. The United States, through the Commercial Space Launch Competitiveness Act of 2015, declared that American citizens and companies have the right to own resources they extract from asteroids and other celestial bodies. Luxembourg, the United Arab Emirates, and Japan have passed similar laws. These nations argue that extracting resources is different from claiming sovereignty – you're not appropriating the asteroid itself, just the materials you remove from it.

This interpretation is contested. Some legal scholars and some nations argue that it violates the spirit of the treaty, that the "province of all mankind" language implies that space resources should benefit everyone, not just those with the capability to extract them.

The disagreement matters. If space resources can be freely appropriated by whoever gets there first, we're back to the pattern of every previous frontier: the powerful capture the wealth, and everyone else is left out. The abundance I've described becomes abundance for some, not abundance for all.

What's needed is an international framework for space resource rights – one that permits development and extraction (which are necessary for space civilization to flourish) while ensuring that the benefits are widely shared (which is necessary for that civilization to be just).

This isn't unprecedented. The Law of the Sea Treaty created a framework for deep seabed mining that balances extraction rights with common heritage principles. It's imperfect and controversial, but it shows that such frameworks can be negotiated.

A space resources framework might include: clear rules about what can be extracted and under what conditions; mechanisms to ensure that some portion of the benefits flows to all nations, not just spacefaring ones; environmental protections to prevent destructive practices; and dispute resolution procedures for conflicts that arise.

The details would be complex. But the principle should be clear: space resources are the common heritage of humanity, and any framework for their development must honor that heritage.

Gap Three: Enforcement is absent.

The Outer Space Treaty has no enforcement mechanism. It relies entirely on the good faith of its signatories.

So far, good faith has been sufficient. Nations have generally complied with the treaty's provisions, not because they fear punishment but because they see compliance as serving their interests. The mutual benefits of a stable, predictable space environment have outweighed the potential gains from violation.

But this equilibrium is fragile. As space becomes more valuable, the temptation to cheat grows stronger. As more actors enter space – nations, companies, perhaps eventually individuals – the chances that someone will defect increase. As the stakes rise, good faith alone may not be enough.

What happens if a nation deploys a weapon in space and claims it's not a weapon? What happens if a company claims a resource that another company believes it has rights to? What happens if a debris-creating event is caused by negligence, and affected parties want compensation?

Currently, there's no clear answer. Disputes would have to be resolved through diplomacy, arbitration, or – in the worst case – conflict. The lack of institutional infrastructure makes every disagreement potentially destabilizing.

The treaty needs teeth. This might include:

A monitoring and verification regime. Nations would report their space activities and submit to inspections, creating transparency that builds trust and detects violations early. This is standard practice in arms control agreements.

A dispute resolution mechanism. An international body – perhaps an extension of existing institutions like the International Court of Justice, or a new specialized tribunal – would hear cases involving space law disputes and issue binding rulings.

Graduated consequences for violations. Rather than relying solely on good faith, the international community would have agreed-upon responses to treaty violations – diplomatic pressure, economic sanctions, exclusion from cooperative programs, and so on.

An implementing organization. A dedicated international agency would oversee treaty compliance, coordinate space activities, manage shared resources, and serve as a forum for ongoing negotiation and adaptation. I'll say more about this in the next chapter.

None of this would be easy to establish. Nations are jealous of their sovereignty and reluctant to submit to international authority. But the alternative – a space environment governed only by power and self-interest – is worse. The question is whether we can build institutions adequate to the challenges we face before those challenges overwhelm us.

Gap Four: Commercial activities are barely addressed.

The treaty was written when space was exclusively the domain of governments. Private companies played supporting roles – building rockets under government contracts, manufacturing satellites to government specifications – but they didn't conduct independent space activities. The idea that private companies might launch their own missions, operate their own space stations, or mine their own asteroids was not on the horizon.

Now it's the horizon.

SpaceX, Blue Origin, Rocket Lab, and dozens of other companies are transforming the space industry. Private space stations are being planned. Commercial lunar landers are being developed. The era of government monopoly is ending.

The treaty addresses this, but barely. Article VI says that nations bear responsibility for their nationals' activities in space and must authorize and supervise those activities. This creates a link between private actions and national accountability. But it says nothing about what rules should govern private activities, what rights private actors have, or how disputes between private parties should be resolved.

The result is a patchwork. Each nation develops its own regulations for its own companies. There's no international consistency, no common standards, no shared framework. A company can shop for the most favorable regulatory environment, incorporating in whichever nation imposes the fewest constraints.

This regulatory arbitrage is already happening. It will intensify as commercial space activities grow. Without international coordination, we'll end up with a race to the bottom – nations competing to attract space companies by offering the loosest oversight.

What's needed is an international framework for commercial space activities – common standards for safety, environmental protection, liability, and resource rights that apply regardless of where a company is incorporated. This doesn't mean heavy-handed regulation that stifles innovation. It means baseline rules that ensure competition happens on a level playing field and that the benefits of commercial space development are not captured entirely by private interests.

Gap Five: The treaty doesn't envision space settlement.

Perhaps the most fundamental gap is conceptual. The treaty was written to govern space exploration and use. It wasn't written to govern space habitation.

Exploration is temporary. You go somewhere, you study it, you come back. Use is instrumental. You put satellites in orbit to serve purposes on Earth. Neither concept captures what happens when people actually live permanently in space – when space becomes not a destination but a home.

The treaty's provisions make sense for exploration and use. But they become awkward when applied to permanent communities.

Consider the prohibition on national appropriation. If a nation cannot claim sovereignty over any part of space, what is the legal status of a habitat built and occupied by that nation's citizens? The habitat itself might be subject to national jurisdiction – like a ship at sea – but what about the space around it? What about the resources it depends on?

Or consider the requirement for international cooperation and consultation. This makes sense when space activities are discrete events – a launch here, a mission there. But permanent communities will engage in continuous activities. Must they consult with all other nations before every action? The framework doesn't scale.

Or consider the principle that space is the province of all mankind. What does this mean for a community that has lived in space for generations, that has developed its own culture and identity, that thinks of a particular habitat as home? Are they merely trustees of humanity's common heritage, or do they have rights of their own?

The treaty doesn't answer these questions because it couldn't imagine them. Space settlement was too remote, too speculative, too far from the urgent concerns of 1967.

Now it's approaching. Not next year, maybe not next decade, but within the lifetime of people alive today. We need to start thinking seriously about the legal and political framework for space communities – communities that may eventually outnumber the population of Earth.

This is perhaps the deepest work that needs to be done: extending the treaty's principles into a domain its framers never envisioned, creating a framework for space civilization rather than merely space activity.

I've described five gaps, but they're really aspects of a single gap: the distance between the treaty's foundational principles and the detailed framework needed to realize them.

The principles are compelling. Space should be peaceful. It should belong to all of humanity. It should be open to everyone. Nations should cooperate. These are, in spirit, the right foundations for the civilization we want to build.

But foundations don't build themselves into structures. The work of construction requires decisions, negotiations, institutions, enforcement. It requires taking general principles and making them specific, taking aspirations and making them operational.

This work is beginning. Various proposals are circulating in diplomatic and academic circles. Some nations are more engaged than others. Progress is slow and uncertain.

But progress is possible. The Outer Space Treaty itself proves that. In 1967, at the height of the Cold War, humanity managed to agree on foundational principles for space. If that was possible then, surely we can build on those principles now.

In the next chapter, I'll describe what a strengthened framework might look like – extending the weapons ban, establishing resource rights, creating enforcement mechanisms, building institutions adequate to the challenges ahead. The gaps are real, but they can be filled. The walls can be built on the foundation. The house can be completed.

Part Four – The Path Forward

☺

Chapter 10: Extending the Ban

Let me tell you about a place where weapons don't exist.

Antarctica is the coldest, driest, most inhospitable continent on Earth. For most of human history, no one lived there. No one could. It was too remote, too brutal, too utterly indifferent to human survival.

When explorers finally reached Antarctica in the early twentieth century, nations began making territorial claims. Britain, Norway, France, Australia, New Zealand, Argentina, Chile – each staked out a slice of the frozen continent, drawing lines on maps, asserting sovereignty over ice and rock and penguin colonies. The claims overlapped. Disputes simmered. The possibility of conflict – even in this most desolate of places – was real.

Then something remarkable happened.

In 1959, twelve nations signed the Antarctic Treaty. They agreed to set aside their territorial claims – not abandon them, but freeze them, holding them in abeyance indefinitely. They agreed that Antarctica would be used exclusively for peaceful purposes. They banned military activities, weapons testing, and nuclear explosions. They established freedom of scientific investigation and required that research results be shared openly.

The treaty has held for over sixty years. Antarctica remains the only continent without war, without armies, without weapons. Scientists from nations that are rivals everywhere else work side by side in research stations. The frozen continent has become a model of international cooperation – proof that humans can, when they choose, set aside competition and create zones of peace.

If we can do it in Antarctica, we can do it in space.

The case for extending the Outer Space Treaty's weapons ban is simple.

The current treaty prohibits weapons of mass destruction in space. This was the urgent concern in 1967 – preventing nuclear weapons from orbiting overhead, ready to strike without warning. The prohibition has held, and the world is safer for it.

But the treaty doesn't clearly prohibit conventional weapons. Anti-satellite missiles. Directed-energy weapons. Kinetic bombardment systems. Electronic warfare platforms. These are not weapons of mass destruction, but they are weapons – designed to damage, destroy, or disable objects in space.

And they're proliferating.

I described the situation in Chapter 6: multiple nations have tested anti-satellite weapons, creating debris fields that endanger everyone. Space forces have been established. Military doctrines now treat space as a "warfighting domain." The taboo against space weapons is eroding with every passing year.

This trajectory leads somewhere dark. If space becomes militarized – if habitats and infrastructure must be defended against attack – then the vision of peaceful abundance becomes impossible. You cannot build a flourishing civilization in a war zone. You cannot plan for generations when destruction might come at any moment. You cannot create the conditions for human thriving when human survival itself is under threat.

The logic is straightforward: if we want space to be the foundation of a better civilization, space must be peaceful. And if space must be peaceful, weapons cannot be permitted there.

The ban on weapons of mass destruction was the right starting point. Now we need to complete what the original treaty began.

What would an extended weapons ban look like?

The core provision would be clear: no weapons in outer space. No nation may place in orbit, install on celestial bodies, or otherwise station in outer space any object designed to damage, destroy, disable, or interfere with spacecraft or space-based infrastructure.

This would cover anti-satellite weapons, whether ground-based, air-launched, or space-based. It would cover directed-energy weapons designed to blind or damage satellites. It would cover kinetic impactors and co-orbital weapons. It would cover electronic warfare systems intended to jam or spoof satellite signals. It would cover any device whose purpose is to harm objects in space.

The ban would not cover war-prevention functions. Reconnaissance satellites, communication satellites, navigation satellites, early warning satellites – these would remain permitted. They don't harm anything; they gather information and transmit signals. Indeed, these functions are stabilizing: they reduce uncertainty, provide warning of attacks, and help prevent the miscalculations that can lead to war.

The distinction between weapons technology and peace technology might seem fuzzy, but it's actually fairly clear. A weapon is designed to cause harm. But a reconnaissance satellite is designed to take pictures, and a communication satellite is designed to relay messages. The purposes are different, and in most cases the hardware is different too.

There would be edge cases, of course. Dual-use technologies exist. A satellite designed for debris removal might theoretically be used to interfere with another nation's spacecraft. A powerful laser designed for space-based manufacturing might theoretically be used as a weapon. These cases would require careful definition and verification protocols.

But edge cases don't invalidate the principle. Every arms control agreement has edge cases. The Chemical Weapons Convention must distinguish between prohibited chemical weapons and permitted industrial chemicals that could theoretically be weaponized. The Biological Weapons Convention must distinguish between prohibited bioweapons research and permitted medical research. Lines can be drawn, even when they're not perfectly sharp.

The goal is not to eliminate every conceivable ambiguity. The goal is to establish a clear norm – weapons don't belong in space – and create mechanisms to enforce it.

The objection I hear most often is that a weapons ban would be unverifiable.

"Space is vast," the argument goes. "Satellites are small. How would you know if a satellite has a hidden weapon aboard? How would you detect a ground-based laser being used for targeting? How would you verify compliance with any confidence?"

This objection sounds serious, but it's less compelling than it appears.

First, verification doesn't require certainty. It requires sufficient confidence that violations would be detected and that the risks of cheating outweigh the benefits. No arms control agreement achieves perfect verification. The question is whether verification is good enough to make the agreement worthwhile.

Second, space is actually more transparent than many domains. Spacecraft are tracked from the ground by multiple nations. Their orbits are known. Their maneuvers are observed. Large-scale deployment of weapons systems would be difficult to hide. A nation attempting to secretly weaponize space would face significant risk of detection.

Third, verification technologies are improving. Space situational awareness – the ability to track and characterize objects in orbit – has advanced dramatically in recent decades. Private companies now offer tracking services that rival government capabilities. The trend is toward more transparency, not less.

Fourth, treaties can include provisions that enhance verification. Nations could be required to notify others before launches, to provide technical specifications of spacecraft, to permit inspections of space hardware before launch. On-orbit inspection capabilities could be developed cooperatively, with agreed-upon protocols. The verification regime could be designed to provide the confidence needed to sustain the agreement.

Fifth, and perhaps most importantly, the alternative to imperfect verification is not perfect security – it's an arms race with no verification at all. If nations refuse to agree on limits because verification is imperfect, they'll end up competing without any constraints. Everyone will be less secure, not more. Imperfect arms control is better than no arms control.

The Antarctic Treaty has held for sixty years without perfect verification. The Outer Space Treaty's ban on weapons of mass destruction has held for nearly as long. Verification is a challenge to be managed, not a reason to abandon the effort.

Another objection: "Nations will never agree to give up their space weapons. The military advantages are too significant. No country will accept constraints that its rivals might violate."

This objection takes the current situation as permanent. It assumes that because nations are developing space weapons today, they will continue developing space weapons forever. It ignores the possibility of change.

But situations do change. Nations that were bitter enemies have become close allies. Arms races that seemed unstoppable have been halted by treaties. Weapons that were once considered essential have been banned and dismantled. History is full of examples of nations choosing cooperation over competition when the conditions were right.

The conditions for a space weapons ban may be more favorable than they appear.

Consider the incentives. Space weapons are expensive to develop and deploy. They create debris that threatens everyone's satellites, including your own. They trigger arms races that leave everyone less secure. They create instabilities that increase the risk of accidental war. The nation that "wins" a space arms race may find that it has won very little – a temporary advantage in a domain now cluttered with debris and fraught with danger.

Consider the asymmetries. Some nations rely more heavily on space assets than others. The United States, for example, depends on satellites for communication, navigation, and reconnaissance more than any other country. This might seem to argue against a weapons ban – why constrain your ability to protect vital assets? But it actually argues for one. In a world where space weapons proliferate, the nation most dependent on space assets is most vulnerable. A mutual ban protects the most exposed party.

Consider the precedents. The Outer Space Treaty itself was negotiated at the height of the Cold War, between superpowers that genuinely believed the other might try to destroy them. If the United States and Soviet Union could agree on the foundations of space law in 1967, today's powers can agree on extending those foundations.

The obstacles are real but not insurmountable. What's needed is leadership – nations willing to propose, negotiate, and champion a weapons-free space. What's needed is public awareness – people who understand the stakes and demand action from their governments. What's needed is the recognition that the alternative to agreement is not the status quo but an accelerating arms race that benefits no one.

Let me be specific about what I'm proposing.

A Protocol to the Outer Space Treaty, negotiated among spacefaring nations and open to all, that would:

Prohibit space weapons. No nation may place in orbit, install on celestial bodies, or otherwise station in outer space any device designed to damage, destroy, disable, or interfere with objects in space. No nation may develop, test, or deploy ground-based, air-based, or sea-based systems designed to damage, destroy, disable, or interfere with objects in space. Weapons, weapons-making, and weapons research are banned from outer space.

Define key terms clearly. The protocol would specify what counts as a weapon, distinguishing prohibited systems from permitted war-prevention and support functions. It would address dual-use technologies with provisions requiring that capabilities beyond certain thresholds be declared and inspected.

Establish verification mechanisms. Nations would be required to provide advance notification of launches, technical specifications of spacecraft, and access to pre-launch inspection. An international monitoring body would track space objects and investigate anomalies. On-orbit inspection capabilities would be developed cooperatively.

Create enforcement provisions. Violations would trigger graduated consequences – diplomatic pressure, economic sanctions, exclusion from cooperative programs, and ultimately, measures authorized by the UN Security Council. The goal would be to make the costs of cheating clearly outweigh any benefits.

Include confidence-building measures. Nations would share information about their space activities, participate in joint exercises, and establish communication channels to prevent misunderstandings. Transparency would reduce suspicion and build the trust necessary for the agreement to hold.

This is not a complete draft treaty. The details would require extensive negotiation among experts in international law, arms control, and space technology. But the outlines are clear enough. We know what a weapons-free space would require. The question is whether we have the will to create it.

I want to address one more objection – the deepest one.

"Even if we ban weapons in space," some argue, "we can't ban weapons on Earth. Nations will still have nuclear arsenals, conventional forces, cyber capabilities. Wars will still be possible. What difference does it make if space is weapons-free when Earth remains armed to the teeth?"

This objection contains a truth: banning space weapons won't create world peace. The conflicts that divide humanity have roots far deeper than any single domain. A weapons-free space won't resolve territorial disputes, ethnic tensions, ideological clashes, or the competition for power that has driven conflict throughout history.

But the objection misses something important.

Space is not just another domain. It's the domain where we might build something new. The resources of space, properly developed, could provide abundance that makes many earthly conflicts obsolete. The blank canvas of space offers the possibility of new social arrangements, new forms of community, new ways of living together. The future of humanity – not just for centuries but potentially for millions of years – will unfold primarily in space, not on Earth.

If space becomes an arena for the same conflicts that have plagued Earth, we'll have squandered the greatest opportunity our species has ever had.

Keeping space weapons-free doesn't solve all of our problems. But it preserves the possibility of solving them. It keeps the door open to the future we want, even as we continue struggling with the present we have.

That's worth doing. That's worth fighting for – through diplomacy, through advocacy, through the slow, difficult work of building international agreement.

Antarctica showed that humans can create zones of peace even in a world of conflict. The original Outer Space Treaty showed that space could be one such zone. Now we need to complete the work – extending the weapons ban, closing the loopholes, creating the verification and enforcement mechanisms that will make the ban stick.

The door is still open. But it won't stay open forever. Every anti-satellite test creates debris that will endanger spacecraft for decades. Every step down the path of weaponization makes it harder to turn back.

The time to act is now. Not because success is guaranteed – it isn't. But because the alternative is to accept, by default, a future we should refuse.

In the next chapter, I'll discuss the institutional framework needed to oversee the weapons ban and the other elements of a strengthened space regime. Rules need enforcers. Principles need institutions. The structure we build must be adequate to the civilization we hope to create.

Chapter 11: The Agency for a Better Cosmos

Rules without referees are suggestions.

You can write the most elegant laws, negotiate the most comprehensive treaties, articulate the most noble principles – and none of it matters if there's no one to interpret the rules, monitor compliance, resolve disputes, and respond to violations. Law without institutions is just words on paper.

The Outer Space Treaty has survived for nearly sixty years with minimal institutional support. There's no International Space Authority, no Orbital Court of Justice, no enforcement body with teeth. The United Nations Office for Outer Space Affairs exists, but it's tiny – a staff of a few dozen people, a budget smaller than many municipal governments, no power to compel anything from anyone.

This institutional thinness was perhaps acceptable when space activities were limited to a few government programs conducting occasional missions. It becomes increasingly inadequate as space becomes crowded, contested, and commercially vibrant. And it will be entirely insufficient for a future in which millions of people live in space, in which the infrastructure of human civilization extends throughout the solar system, in which the stakes of getting space governance wrong are measured in the fate of our species.

We need institutions adequate to the challenges we face. We need, to give it a name, an Agency for a Better Cosmos.

What would such an agency do?

Let me describe its functions, then discuss how it might be structured and governed.

Function One: Monitoring and Verification

The agency would track what's happening in space. It would maintain comprehensive awareness of all objects in orbit – satellites, debris, spacecraft in transit. It would receive notifications of planned launches and maneuvers. It would operate or coordinate a network of ground-based and space-based sensors.

It would know, to the greatest extent possible, what's up there and what it's doing.

This function is partly technical and partly diplomatic. The technical side involves sensors, databases, analysis tools, and the expertise to use them. The diplomatic side involves agreements with nations to share information, protocols for notification and transparency, and mechanisms to investigate when something unexpected occurs.

Monitoring serves multiple purposes. It enables verification of arms control agreements – detecting potential weapons deployments, tracking suspicious maneuvers, providing early warning of treaty violations. It supports space traffic management – preventing collisions, coordinating orbits, managing the increasingly crowded environment of near-Earth space. It creates accountability – making it harder for any actor to engage in harmful behavior without detection.

The agency wouldn't need to do all this alone. National space agencies, private tracking companies, academic institutions – many organizations contribute to space situational awareness. The agency's role would be to coordinate, integrate, and ensure that the collective picture is comprehensive and shared.

Function Two: Regulation and Standards

The agency would develop and administer rules for space activities. Safety standards for spacecraft design. Environmental requirements for debris mitigation. Protocols for orbital maneuvers and collision avoidance. Licensing procedures for resource extraction. Spectrum allocation for communications. Planetary protection guidelines.

Some of these rules exist already, developed by various national and international bodies. But they're fragmented, inconsistent, and often voluntary. A company operating from one nation follows different rules than a company operating from another. Standards that work well on paper aren't always followed in practice.

The agency would create a unified regulatory framework – not necessarily replacing national regulations but establishing baseline international standards that all actors must meet. This would level the playing field, preventing regulatory arbitrage. It would provide clarity for companies making long-term investments. It would ensure that the activities of one actor don't impose unacceptable risks on others.

Regulation sounds bureaucratic, even stifling. But good regulation enables rather than constrains. It creates the predictable environment in which investment and innovation can flourish. It prevents the tragedies of the commons that occur when everyone pursues their own interest without regard for collective consequences. It's the difference between a freeway with traffic laws and a demolition derby.

Function Three: Dispute Resolution

When conflicts arise – and they will arise – the agency would provide mechanisms to resolve them.

A company believes another company has infringed on its resource rights. Two nations disagree about whether a particular activity complies with the weapons ban. A habitat suffers damage from debris created by another actor's negligence. A spacecraft maneuver is alleged to have violated safety protocols.

Currently, such disputes have no clear resolution path. They might be addressed through bilateral diplomacy, through existing international courts with uncertain jurisdiction, through economic pressure, or not at all. The lack of reliable dispute resolution creates uncertainty, discourages investment, and increases the risk that conflicts will escalate.

The agency would include a tribunal – a judicial body with expertise in space law, authorized to hear cases, make findings, and issue binding rulings. Parties would agree, by participating in the treaty framework, to accept the tribunal's jurisdiction. Decisions would be enforceable through the mechanisms available to the international community.

This wouldn't resolve every conflict. Conflict management, as distinguished from conflict resolution, has a role to play: Some disputes involve fundamental interests that no tribunal can adjudicate. But it would provide a peaceful path for the many conflicts that arise from ambiguity, misunderstanding, or competing interpretations of rules. It would channel disagreements into legal processes rather than power struggles.

Function Four: Enforcement

Rules need consequences. When actors violate the treaty framework – deploying prohibited weapons, extracting resources without authorization, creating debris through negligence, refusing to comply with tribunal rulings – something must happen.

Enforcement in international law is always difficult. There's no world government, no global police force, no authority that can simply compel compliance. International institutions work through cooperation, persuasion, and the coordinated action of member states.

But coordinated action can be powerful. Economic sanctions imposed by major economies can cripple violators. Exclusion from cooperative programs can deny access to valuable resources and opportunities. Diplomatic isolation can impose political costs. In extreme cases, more coercive measures – authorized by appropriate international bodies – might be necessary.

The agency would coordinate these responses. It would investigate alleged violations, make determinations about compliance, recommend appropriate responses, and facilitate collective action by member states. It wouldn't have independent coercive power – that would require a transformation in international relations that isn't imminent – but it would provide the framework within which enforcement happens.

Function Five: Coordination and Facilitation

Beyond regulation and enforcement, the agency would serve as a forum for ongoing cooperation.

Space activities require coordination. Orbits must be deconflicted. Communications frequencies must be allocated. Rescue capabilities must be maintained. Scientific research benefits from collaboration. Technology development can be shared. Information about hazards should flow freely.

The agency would convene working groups, host conferences, facilitate agreements, and provide the institutional infrastructure for international cooperation on space matters. It would be a meeting place where technical experts, diplomats, industry representatives, and civil society could address common challenges.

This facilitation function might seem less dramatic than regulation or enforcement, but it could be equally important. Many problems in space governance arise not from bad faith but from lack of coordination. Actors pursuing their own interests, without mechanisms to harmonize those interests, create outcomes that benefit no one. The agency would provide those mechanisms.

How would such an agency be governed?

This is where things get difficult. International institutions reflect the power dynamics of their members. The United Nations Security Council, for example, gives veto power to five permanent members – a structure that made sense in 1945 but now often paralyzes the body. The International Monetary Fund weights voting power by financial contribution, privileging wealthy nations. Every governance structure embodies choices about who has power and how it's exercised.

For the proposed Agency for a Better Cosmos, several principles should guide the design.

Broad participation. The agency should include all nations willing to accept its framework, not just the current spacefaring powers. Space is the province of all mankind; its governance should reflect that. Small nations and large, wealthy and poor, technologically advanced and developing – all should have voice and vote.

Expertise matters. Space governance involves technical questions that require specialized knowledge. The agency should incorporate expertise – scientists, engineers, legal scholars – in ways that inform decision-making without supplanting political accountability.

Weighted responsibility. While all nations should participate, those whose activities have greater impact might reasonably bear greater responsibility – and perhaps have greater say in relevant decisions. The largest spacefaring nations shouldn't be able to be outvoted on matters affecting their vital interests by coalitions of nations with minimal space presence. Finding the right balance between equality and proportionality is essential.

Evolution over time. The distribution of space capabilities will change. Nations that are minor players today may become major players tomorrow. The governance structure should be able to adapt, shifting influence as realities shift, rather than locking in a particular moment's power distribution.

Accountability and transparency. The agency's decisions should be visible and explicable. Its processes should be open to scrutiny. Its officials should be accountable for their actions. Legitimacy depends on the perception that the agency serves the common good rather than particular interests.

No governance structure will satisfy everyone. Any design involves tradeoffs, and different nations will prefer different balances. The negotiation of the agency's structure would be contentious, perhaps as contentious as the negotiation of its substantive provisions.

But contentious negotiations can succeed. The United Nations itself, the World Trade Organization, the International Criminal Court – all were products of difficult negotiations that somehow reached conclusion. The key is maintaining focus on the shared interest in having a functioning institution, even while disagreeing about its precise design.

I can imagine the skeptics' response.

"This sounds like world government. Nations will never agree to cede sovereignty to an international bureaucracy. The whole proposal is utopian fantasy."

Let me address this directly.

First, the agency I'm describing is not world government. It would have limited, specific functions related to space activities. It wouldn't tax anyone, conscript anyone, or govern anyone's daily life. It would be more like the International Civil Aviation Organization or the International Telecommunication Union –

specialized bodies that coordinate activities in particular domains – than like a global sovereign.

Second, nations cede sovereignty to international institutions all the time. They do it when the benefits of coordination outweigh the costs of constraint. They do it when unilateral action produces worse outcomes than collective action. They do it when the alternative to agreement is chaos.

The World Trade Organization constrains national trade policy. The International Monetary Fund imposes conditions on borrowing nations. The International Criminal Court can prosecute citizens of member states. The Non-Proliferation Treaty subjects nations to intrusive inspections of their nuclear facilities. None of these institutions represents "world government," but all involve nations accepting international authority over matters once considered purely domestic.

Space governance follows the same logic. As space becomes more important, the costs of uncoordinated action rise. Debris accumulates. Orbits become congested. Resources are contested. The risk of conflict grows. At some point, the benefits of coordination become compelling enough that nations accept the constraints necessary to achieve them.

We're approaching that point. Perhaps we've reached it already. The question is not whether international coordination of space activities will happen, but what form it will take – whether it will be deliberate and well-designed or haphazard and inadequate.

Third, the agency would not spring into existence fully formed. It would start small, perhaps as an expansion of existing UN bodies, with limited functions and modest authority. Over time, as it demonstrated competence and earned trust, its role could expand. Institution-building is incremental. It proceeds step by step, each step creating the conditions for the next.

The Outer Space Treaty itself followed this pattern. It established principles; later agreements added detail. The treaty framework has grown and evolved over decades. A new agency would be part of that evolution, not a revolutionary break with it.

Let me paint a picture of how the agency might work in practice.

It's 2045. The Agency for a Better Cosmos – ABC for short – has been operating for a decade. Its headquarters is in Vienna, near the existing UN space offices, but it has regional centers in Houston, Beijing, Bangalore, and Nairobi.

A company based in Luxembourg has begun preliminary operations at an asteroid it discovered and registered three years ago. A company based in Japan claims that the asteroid's orbital neighborhood was already designated for its own future operations and that the Luxembourg company is infringing on its rights.

Both companies file claims with ABC's Resource Tribunal. Lawyers present arguments; technical experts testify about orbital dynamics and resource distribution; the tribunal reviews the registration records and the applicable regulations. After deliberation, it issues a ruling: the Luxembourg company may continue operations on the asteroid itself, but must avoid certain maneuvers that would interfere with the orbital corridors the Japanese company had previously registered. Both parties accept the ruling, adjust their operations, and continue their work.

Meanwhile, ABC's monitoring division has detected an anomaly – a satellite launched by a non-member state has maneuvered close to a communications satellite operated by a member state. The approach doesn't violate any clear rule, but it's unusual and concerning. ABC's verification team requests an explanation through diplomatic channels. The non-member state provides technical data showing that the maneuver was for debris avoidance – a piece of space junk required an unexpected orbit change. The explanation is plausible; the data supports it; the matter is closed but logged for future reference.

In a conference room in Bangalore, a working group is finalizing new guidelines for habitat construction standards. Engineers from six countries have spent months developing consensus recommendations for structural integrity, life support redundancy, and evacuation protocols. The guidelines will be submitted to ABC's regulatory council for adoption. Once adopted, they'll become binding on all member states and their licensed operators.

None of this is dramatic. It's bureaucracy – meetings and filings and procedures. But this kind of bureaucracy is what makes complex systems function. It's what allows diverse actors with different interests to coexist and cooperate. It's what turns principles into practice.

I want to be honest about the difficulties.

Creating an Agency for a Better Cosmos would require sustained diplomatic effort over many years. It would require the major spacefaring powers – the United States, China, Russia, Europe, India, Japan, and others – to agree on fundamental matters where their interests diverge. It would require smaller nations to accept an institution they didn't design. It would require compromises that leave everyone somewhat dissatisfied.

The current trajectory is not toward such an institution. If anything, it's away from it – toward fragmentation, ultracompetition, and the erosion of existing frameworks. The creation of national space forces, the proliferation of uncoordinated commercial activities, the testing of anti-satellite weapons – these developments suggest that nations are preparing for zero-sum competition, not win-win cooperation.

But trajectories can change. The current moment is not destiny. History is full of examples where ultracompetitive dynamics reversed, where nations that were arming against each other chose instead to negotiate limits. Sometimes it takes a crisis – a near-catastrophe that focuses minds on the costs of continued zero-sum or lose-lose competition. Sometimes it takes leadership – individuals and governments that articulate a different vision and mobilize support for it.

I don't know what will trigger the shift toward serious space governance. But I believe the shift is necessary and possible. The alternative – an increasingly chaotic and contested space environment, an arms race extending into orbit, the squandering of humanity's greatest opportunity – is not acceptable.

We need institutions adequate to the challenges we face. We need an Agency for a Better Cosmos, or something like it. And we need to start building it now, before the window closes.

Chapter 12: Building the New Civilization

We've spent several chapters discussing problems – military threats, commercial capture, institutional gaps, the doors that are closing. Now let's open a different door. Let's walk through it and look around.

What does the civilization we're trying to build actually look like?

I want to make this concrete. Not science fiction, not utopian fantasy, but a realistic picture of what becomes possible when the principles we've discussed are actually implemented. When space is genuinely peaceful. When abundance is genuinely shared. When the blank canvas is painted with intention and care.

Close your eyes for a moment. Imagine it's a century from now. The framework we've been discussing – the extended weapons ban, the resource-sharing agreements, the Agency for a Better Cosmos – has been in place for generations. What do you see?

You see communities.

Not space stations – those cramped, utilitarian outposts where specialists conduct temporary missions. Communities. Places where people are born, grow up, fall in love, raise children, and live abundantly. Places with schools and parks and restaurants and libraries. Places with neighborhoods and traditions and local politics and Friday night gatherings.

The first GIF habitats were built in the twenty-first century. They were dependent on Earth for supplies and expertise. But they proved the concept. People could live in space, really live there, for entire lifetimes.

By the twenty-second century, the experiments had become a civilization. Hundreds of habitats orbited in the stable regions near Earth, at the Lagrange points, in the asteroid belt, around Mars and beyond. Each habitat was a world unto itself – a cylinder miles long, rotating gently, its interior landscaped with forests and fields and villages.

Some habitats were urban, densely populated, buzzing with commerce and culture. Others were rural, agrarian, deliberately simple. Some were organized around particular industries – manufacturing, research, the arts. Others were intentional communities built around shared values – religious, philosophical, political. Some welcomed newcomers; others were more insular. Some were governed democratically; others experimented with different arrangements.

The diversity is hard to overstate. On Earth, geography constrains community formation. You can't easily relocate to a society that better fits your values; you're stuck with your neighbors, your nation, your inherited circumstances. In space, communities are chosen. If you don't like how your habitat is governed, you can leave. If you dream of a different way of life, you can find others who share that dream and build it together.

This doesn't mean conflict has disappeared. People still disagree, still compete, still struggle for status and meaning. Human nature hasn't changed. But the context has changed. When there's room enough for everyone, when resources aren't scarce, when exit is always an option, the conflicts that remain are more manageable. They're disagreements within a framework of abundance rather than battles for survival.

You see abundance.

Not luxury for the few – abundance for all. Every person in space civilization has access to the basics of a good life: food, shelter, healthcare, education, opportunity. Not because of charity, and not because of coercion, but because the systems produce enough for everyone and the frameworks ensure it's distributed.

The abundance machine described in Chapter 3 became real. Solar collectors spanning millions of square kilometers harvest the endless energy of the sun. Automated mining operations extract materials from asteroids and process them into useful forms. Manufacturing facilities – some crewed, some entirely robotic – produce everything from spacecraft to furniture to works of art. The productive capacity is so vast that scarcity, in the traditional sense, has become obsolete.

This required decisions. The abundance didn't distribute itself. Early in space development, there were fierce debates about resource rights, about who would benefit from the wealth of the solar system. The framework that emerged – building on the Outer Space Treaty's principle that space is the province of all mankind – established that certain resources would be held in common and their benefits shared widely.

Different habitats implemented this differently. Some created universal basic incomes funded by resource revenues. Others provided goods and services directly. Still others used more complex arrangements involving partial ownership, cooperatives, and novel economic structures that had no precedent on Earth. The common thread was a commitment to ensuring that no one fell through the cracks, that the floor of human existence was high enough for dignity.

Above that floor, people still strove. They still competed, created, sought excellence and recognition. The end of scarcity didn't mean the end of ambition. It meant that ambition could be directed toward creation rather than mere survival, toward meaning rather than desperation.

You see peace.

The weapons ban held. Space became, and remained, a weapons-free zone. The Agency for a Better Cosmos monitored, verified, and occasionally intervened when violations threatened. But violations were rare, because the incentives pointed toward compliance. In a civilization of abundance, the gains from aggression were minimal and the costs were high. War made no sense when there was nothing worth fighting over.

This didn't mean the end of all conflict. Disputes still arose – between individuals, between communities, between the occasional habitat that pushed boundaries. But these disputes were channeled through institutions: courts, arbitration panels, diplomatic processes. Violence became aberrant rather than normal, a breakdown of systems rather than a feature of them.

On Earth, the peace dividend was transformative. With space resources flowing and space-based energy beaming down, the material pressures that had driven so much conflict eased. Nations that had armed against each other found less reason to do so. The vast sums once spent on weapons could be redirected to other purposes. The change didn't happen overnight – old hatreds and old institutions had momentum – but over generations, the texture of international relations shifted.

I don't want to paint this too rosily. Humanity didn't become angels. Politics remained contentious. Some communities made bad choices; some experiments failed; some individuals caused harm. The problems of human existence – the search for meaning and purpose – persisted. Peace and abundance solved some problems but not all problems.

But consider what was avoided. The nuclear war that had threatened for a century never came. The resource wars that many predicted as Earth's supplies dwindled never materialized. The climate catastrophe that loomed was mitigated by space-based energy and the migration of heavy industry off-planet. The future that had seemed so threatening became, instead, an opening.

You see freedom.

Not the cramped freedom of choosing between limited options, but genuine freedom – the freedom that comes when material constraints fall away and social constraints are chosen rather than imposed.

In space civilization, people have options. If you're born in a habitat whose values don't suit you, you can leave. If you want to start a new community organized around different principles, you can gather like-minded people and do so. If you want to live alone in a small habitat at the edge of the solar system, that's possible too. The coercion that arises from having nowhere else to go, no other options, no exit – that coercion diminishes when the universe is open.

This freedom extended to ways of life that would be impossible on Earth. Low-gravity habitats for those who preferred it. Habitats with unusual day-night cycles, unusual climates, unusual social arrangements. Communities organized around art, around contemplation, around adventure, around whatever purposes people found meaningful. The diversity of human aspiration, no longer squeezed into the limited molds that Earth's geography and history had created, flourished.

Freedom also meant freedom from the old hierarchies. The social structures of Earth – the accumulated inequalities, the entrenched privileges, the class systems hardened over centuries – didn't automatically transfer to space. New communities could start fresh, without landlords whose ancestors had seized the land generations ago, without aristocracies whose position rested on ancient conquests, without the calcified social orders that make change so difficult on Earth.

This didn't mean perfect equality. Talent, effort, luck – these still created differences. Some people were more influential than others, more wealthy, more admired. But the differences were less extreme and less permanent. Mobility was real. The barriers that kept people in their place on Earth were, in space, largely absent.

You see Earth, cherished and healing.

One of the great fears about space development was that it would lead to abandoning Earth – that the privileged would escape to orbital paradises while the planet below descended into ruin. This fear was not irrational; the history of colonization offered plenty of examples of exploitation and abandonment.

But that's not how it unfolded.

Space civilization remained connected to Earth. Most people in space had ancestors on Earth; many had relatives still living there. The cultural, emotional, and economic ties were deep. Earth was not a used-up husk to be discarded but the birthplace of everything humanity had ever been – the repository of history, the cradle of life, irreplaceable and beloved.

More practically, the flow of resources went both ways. Space-based solar power beamed clean energy to Earth's surface. Materials extracted from asteroids reduced the pressure to mine Earth's diminishing reserves. Heavy industries that had polluted Earth's air and water relocated to space, where their byproducts could be managed without damaging living systems.

Earth's population, freed from the pressure of supporting all human activity, stabilized and eventually declined. Not through catastrophe but through choice – people had other options now, other places to build lives. The areas that had been stripped for resources began to recover. Forests returned to denuded hillsides. Rivers ran clear again. Species that had been on the brink of extinction found space to survive.

This wasn't paradise. Earth still had problems, still had inequality, still had the complex politics of billions of people sharing a finite planet. But the trajectory had changed. Instead of a planet being exhausted by demands it couldn't meet, Earth became one part of a larger civilization – honored, protected, allowed to heal.

You see purpose.

This might be the most important thing.

The great risk of abundance is meaninglessness. When survival no longer requires struggle, when material needs are met automatically, what do people do? What gets them out of bed in the morning? What gives their lives direction and significance?

This question haunted early space civilization. Some people flourished in the new conditions; others floundered. The old purposes – working to feed your family, building a career, accumulating enough security to face an uncertain future – lost their urgency. Not everyone found new purposes to replace them.

But most people did. They found purpose in creation – art, music, literature, the endless human drive to make something new. They found purpose in discovery – science, exploration, the investigation of a universe that remained vastly mysterious. They found purpose in connection – family, friendship, community, the web of relationships that gives human life its texture. They found purpose in service – teaching, healing, helping, the satisfaction of contributing to others' wellbeing. They found purpose in cultivation – gardens, ecosystems, the tending of living things. They found purpose in mastery – skills, crafts, the deep engagement with challenging activities. They found purpose in transcendence – contemplation, spirituality, the search for meaning beyond the material.

Abundance didn't eliminate the need for purpose. It liberated it. When you don't have to spend your life struggling to survive, you can spend it on whatever you find most meaningful. The range of human purposes, no longer constrained by necessity, expanded in directions that would have been impossible when scarcity ruled.

I've painted an optimistic picture. Too optimistic, perhaps. Let me add some shadows.

The transition wasn't smooth. There were crises, setbacks, moments when the whole project seemed on the verge of collapse. The first habitat failure – a catastrophic structural breach that killed everyone aboard – shocked space civilization and led to years of recrimination and revised safety standards. Economic dislocations as Earth's industries lost their monopolies created hardship and resentment. Political conflicts between Earth-based nations and space communities flared repeatedly before new equilibria were found.

Not everyone benefited equally. Despite the frameworks for sharing, some communities accumulated more than others. Some individuals fell through the cracks of even the most generous systems. The promise of abundance for all was imperfectly kept.

Some communities made terrible choices. Cults formed in isolated habitats, with charismatic leaders exploiting the freedom from outside oversight. Experiments in social organization sometimes produced misery rather than flourishing. The blank canvas allowed for painting bad pictures as well as good ones.

And the old problems persisted in new forms. People still hurt each other. Relationships still failed. Mental illness still tormented. The human condition, even in abundance, remained the human condition – shot through with suffering as well as joy.

But here's what matters: the trajectory was upward.

Not a straight line – there were dips and reversals. Not utopia – problems remained and new ones emerged. But on the whole, over time, across generations, things got better. More people lived lives of dignity. More of human potential was realized. More of the darkness that had shadowed human history – the poverty, the violence, the oppression, the despair – was pushed back.

This is what we're working toward. Not perfection but progress. Not the elimination of all problems but the creation of conditions in which problems can be addressed rather than festering. Not a final state but an opening – a door into a future where human flourishing is possible on a scale we can barely imagine today.

The civilization I've described is not inevitable. It's not even probable, given current trajectories. It's possible – possible if we make the right choices, build the right frameworks, establish the right institutions. Possible if we extend the Outer Space Treaty, ban weapons, create mechanisms for sharing resources, build the Agency for a Better Cosmos.

That's why the work described in this book matters. We're not just negotiating treaties and designing institutions. We're shaping the future. The decisions we make in the next few decades will determine whether the picture I've painted becomes reality or remains a dream.

In the next chapter, we'll return to the present and face squarely the obstacles in our path. The vision is clear; the question is whether we can reach it. What might stop us? What might we do about it?

The picture I've painted is beautiful. Now we have to earn it.

Part Five – The Choice Before Us

☺

Chapter 13: Why We Might Fail

I've spent most of this book making a case for hope. Now let me make a case for worry.

The vision I've described – peaceful, abundant space civilization governed by fair institutions – is possible. The physics works. The engineering is plausible. The principles are sound. But possible is not the same as likely. Between here and there lies a minefield of obstacles, any one of which could derail us.

If we're going to succeed, we need to understand clearly why we might fail. Not to induce despair, but to know where to focus our efforts. A doctor who doesn't understand a disease can't treat it. An engineer who doesn't understand failure modes can't prevent them. We need to look at the obstacles with clear eyes.

Here, then, are the reasons we might not make it.

The Great Powers Don't Want It

The most fundamental obstacle is that the nations with the most power to shape space development may not want the future I've described.

The United States, China, and Russia are the dominant spacefaring powers. Their decisions will largely determine what happens in space over the coming decades. And all three are currently moving in directions that undermine the vision of peaceful, shared abundance.

The United States has established a Space Force and explicitly treats space as a warfighting domain. Its national security strategy emphasizes maintaining space superiority – the ability to deny adversaries the benefits of space while preserving those benefits for itself. The U.S. has resisted international agreements that would constrain its freedom of action in space, viewing them as potentially limiting its military advantages.

China is rapidly expanding its space capabilities, with an explicit goal of becoming a "space power" to rival the United States. It has tested anti-satellite weapons, developed counter-space capabilities, and shown little interest in arms control agreements that might limit its rise. Its approach to space resources emphasizes national development rather than international sharing.

Russia, though its space program has declined from Soviet-era heights, retains significant capabilities and has demonstrated willingness to use them aggressively. Its anti-satellite tests, its doctrine of asymmetric warfare (unconventional strategies), and its general hostility to Western-led international institutions make it an unlikely partner for the cooperative frameworks I've described.

These three powers have competing interests, deep mutual suspicion, and a history of treating international agreements as constraints to be avoided or exploited rather than frameworks to be honored. Getting them to agree on an extended weapons ban, a resource-sharing regime, and a robust international agency would require a transformation in their relationships that nothing in current trends suggests is coming.

This is not a peripheral problem. It's the central problem. Without the great powers on board, the institutional framework I've described cannot be built.

Commercial Interests Oppose Constraints

The private space industry is booming. SpaceX, Blue Origin, Rocket Lab, and dozens of smaller companies are transforming what's possible in space. They're driving down launch costs, enabling new applications, and bringing entrepreneurial energy to a domain long dominated by government bureaucracies.

This is genuinely good. Private enterprise has been essential to space development, and it will continue to be essential. The future I've described depends on commercial capabilities that government programs alone could never deliver.

But commercial interests are not automatically aligned with the common good.

Companies want to maximize profits. They want freedom to operate without regulatory constraints. They want property rights that are clear, enforceable, and favorable to their interests. They want rules of the game that protect their investments and exclude competitors.

The framework I've described – with its resource-sharing provisions, its robust international oversight, its constraints on unilateral action – cuts against these preferences. Companies that have invested billions in developing space capabilities don't want to share the returns with nations that contributed nothing. They don't want international bureaucrats second-guessing their operations. They don't want to be bound by agreements negotiated by governments with different priorities.

The lobbying power of commercial space interests is growing rapidly. In the United States, space companies have become significant political players, funding campaigns, shaping legislation, and influencing regulatory agencies. Their voice in policy debates is loud and well-funded.

The voice of "all of humanity" – the diffuse, unorganized interest in space being developed for everyone's benefit – has no comparable representation. There's no lobby for the common heritage of mankind. There's no PAC for future generations. There's no industry association for people who don't yet exist.

This asymmetry means that commercial interests will be overrepresented in the decisions that shape space governance. The rules that emerge will reflect their preferences more than they reflect broader human interests. The capture of space development by private wealth, which I warned about in earlier chapters, is not a distant threat. It's happening now, in regulatory proceedings and legislative chambers where most people aren't paying attention.

The Public Doesn't Care

Here's a hard truth: most people don't think about space governance at all.

They might enjoy science fiction movies. They might find rocket launches exciting. They might have vague positive feelings about astronauts and space exploration. But the details of space policy – treaty frameworks, resource rights, institutional design – are invisible to them. They don't know about these issues, don't understand why they matter, and don't demand action from their governments.

This is not because people are stupid or shallow. It's because people have limited attention, and space governance doesn't seem urgent compared to the issues that affect their daily lives. Inflation, healthcare, crime, immigration, climate change – these problems are concrete and immediate. The Outer Space Treaty is abstract and remote.

Politicians respond to what voters care about. If voters don't care about space governance, politicians won't prioritize it. The diplomats and experts who work on these issues will be marginalized, underfunded, ignored. The international negotiations that could produce better frameworks will languish.

Public apathy creates a vacuum that special interests fill. When ordinary citizens aren't watching, the decisions get made by whoever is watching – which means commercial interests, military planners, and bureaucratic actors pursuing their own agendas. The common good, unrepresented in these deliberations, gets short shrift.

Changing this would require a massive effort in public education and mobilization. People would need to understand what's at stake, connect it to their own interests, and demand action. This has happened on other issues – environmental protection, civil rights, public health – but it requires sustained effort over years or decades.

On space governance, that effort has barely begun.

International Cooperation Is Failing Everywhere

Even if the great powers wanted to cooperate, even if commercial interests were willing to accept constraints, even if the public demanded action – the mechanisms for international cooperation are in disrepair.

Look around the world. The international institutions built after World War II are straining under pressures they were never designed to handle. The United Nations is gridlocked on major issues. The World Trade Organization's dispute resolution system has been paralyzed. Arms control agreements are collapsing. Climate negotiations produce agreements that are too weak and poorly enforced.

The reasons are complex: rising nationalism, populist backlash against global elites, the decline of American hegemony, the rise of China, the erosion of shared norms, the corrosive effects of social media on public discourse. Whatever the causes, the result is clear. International cooperation is harder now than it was a generation ago, and the trend is getting worse, not better.

Creating the framework I've described – an extended weapons ban with verification, a resource-sharing regime, a robust international agency – would require international cooperation at a level that seems almost unimaginable in the current environment. It would require sustained negotiations over years, compromise by all parties, and implementation mechanisms that actually work.

If we can't even cooperate effectively on climate change – a clear and present danger to everyone – how can we expect to cooperate on space governance, whose benefits are more diffuse and distant?

Technology Outruns Governance

There's a pattern in human affairs: technology moves fast; governance moves slow.

New capabilities emerge before we've figured out how to manage them. By the time rules are established, the situation has changed. The regulations designed for one era are applied awkwardly to another. We're always playing catch-up, always a step behind.

Space is no exception. The Outer Space Treaty was written for an era of government-dominated space programs conducting occasional missions. It didn't anticipate mega-constellations of thousands of satellites, or commercial space stations, or asteroid mining, or space tourism, or AI-controlled spacecraft, or any of the developments that are now upon us.

The pace of change is accelerating. What's possible in space is expanding faster than our ability to understand it, much less govern it. By the time we negotiate a treaty on asteroid mining, the first asteroids may already have been claimed. By the time we establish rules for habitat construction, the first habitats may already be built – by actors who set their own rules.

This dynamic favors those who move fast over those who deliberate carefully. It favors unilateral action over multilateral agreement. It favors facts on the ground over principles in treaties. The institutions I've described require consensus-building processes that take time – time that technology doesn't give us.

The window of opportunity I've discussed throughout this book is not just about political will. It's also about timing. The foundational patterns of space civilization are being established now, through actions that won't wait for governance structures to catch up.

We Might Just Not Be Smart Enough

This is the hardest obstacle to discuss, because there's nothing obvious to do about it.

The challenges of space governance are immensely complex. They involve physics, engineering, economics, law, politics, diplomacy, and human psychology. They span timescales from immediate to centuries-long. They require anticipating developments that no one can foresee, designing institutions that will remain effective as circumstances change, and aligning the interests of actors with wildly different values and goals.

Humans have solved complex problems before. But we've also failed to solve them. We've built civilizations and watched them collapse. We've created institutions and watched them corrode. We've achieved coordination and watched it unravel.

There's no guarantee that we're capable of the kind of sustained, intelligent, cooperative effort that building a good space civilization requires. Our cognitive biases, our tribal instincts, our short-term thinking, our difficulty grasping large numbers and long timeframes – all of these work against us.

Maybe we're just not up to the task. Maybe the vision I've described, while theoretically possible, exceeds the actual capabilities of the human species. Maybe we're the kind of creature that can get to space but can't govern ourselves once we're there.

I don't believe this. I think humans are capable of extraordinary things when conditions align. But it would be dishonest to pretend that success is assured. The track record is mixed, and the challenges we face now are greater than any we've faced before.

The Worst-Case Scenario

Let me describe where we end up if we fail.

Space is militarized. Weapons platforms orbit overhead. Every major power has anti-satellite capabilities and space-based attack systems. The threat of space conflict hangs over international relations, adding another dimension of danger to an already perilous world.

Space resources are captured by first movers. A handful of corporations and nations control the asteroids and the orbital positions. They extract wealth on a staggering scale, but that wealth flows to shareholders and sovereign wealth funds, not to humanity at large. Most of Earth's population sees no benefit from the abundance of space.

Habitats exist, but they're enclaves of the privileged. The rich live in orbital paradises; everyone else looks up from a depleted planet. The dream of abundant space civilization becomes a bitter joke – abundant for some, inaccessible for most.

Earth continues to struggle. Without the relief that shared space resources could have provided, pressures like climate change, resource depletion, population, or migration continue to build. Conflicts intensify. The hope that space offered a way out becomes another hope betrayed.

The blank canvas is painted, but with the same old picture – domination, extraction, conflict, inequality. Humanity's greatest opportunity is squandered because we couldn't get our act together in time.

This is not inevitable. But it's where current trajectories seem to lead. It's what happens if we do nothing, if we let things drift, if we assume that good outcomes will emerge automatically.

I've spent this chapter dwelling on darkness. Not because I think failure is certain, but because I think clarity about risks is necessary for taking them seriously.

In the next chapter, I'll turn to reasons for hope. Because despite everything I've said, I do have hope. The obstacles are real, but they're not insurmountable. The path is difficult, but it's not impossible. And the stakes are high enough to justify extraordinary effort.

We might fail. But we might not. And everything depends on which of those possibilities we make real.

Chapter 14: Why We Might Succeed

I've just given you every reason to despair. Now let me give you reasons not to.

The obstacles are real. I won't pretend otherwise. Great power competition, commercial capture, public apathy, institutional decay, the relentless pace of technology – these are serious challenges. Anyone who promises easy solutions is selling something.

But obstacles are not the same as impossibilities. Throughout human history, people have faced daunting challenges and sometimes – not always, but sometimes – they've overcome them. The question is not whether success is guaranteed but whether it's possible. And whether, being possible, it's worth pursuing.

I believe it is. Here's why.

The Treaty Already Exists

We're not starting from nothing.

The Outer Space Treaty has been in force for nearly sixty years. Over one hundred nations have ratified it. Its core principles – space as the province of all mankind, peaceful purposes, no national appropriation – are embedded in international law. Every spacefaring nation has accepted these principles, at least formally.

This matters more than it might seem.

When you're trying to build something new, the hardest part is often establishing the foundation. Getting nations to agree on basic principles, creating the initial framework, achieving the first consensus – these are the most difficult steps. Once a foundation exists, building on it is comparatively easier.

We have the foundation. The principles we need are already agreed. What we're proposing is not a revolutionary new framework but an extension and strengthening of an existing one. We're filling gaps, not starting over.

This doesn't make success easy. But it makes it less hard. Every diplomat who works on space governance doesn't have to begin by justifying why space should be peaceful or why it should benefit all of humanity. Those arguments have been won. The Outer Space Treaty won them, decades ago.

The skeptic might say: "But the treaty is being eroded. Nations are pushing against its limits. The principles exist on paper but not in practice."

True enough. But erosion can be reversed. Principles that exist on paper can be reinforced in practice. The treaty provides a rallying point, a shared reference, a legitimate basis for demanding that nations live up to their commitments. Without the treaty, we'd have to create that basis from scratch. With the treaty, we can invoke what nations have already agreed to.

The foundation is cracked but not broken. Our job is to repair and build on it, not to lay a new one.

Mutual Interest Is Real

The obstacles I described in the last chapter are real, but so is the mutual interest in overcoming them.

Consider the great powers. Yes, they're competing. Yes, they're developing space weapons. Yes, they're suspicious of agreements that might constrain their freedom of action. But they also share interests that cooperation would serve.

All of them depend on space assets. Satellites for communication, navigation, reconnaissance, early warning – these are essential to contemporary security infrastructures and modern economies. All of them are vulnerable to attacks on those assets. An arms race in space makes everyone less secure, including the nations currently leading the race.

All of them face the debris problem. Every anti-satellite test creates thousands of fragments that threaten everyone's spacecraft. The Kessler syndrome – a cascade of collisions that renders entire orbital regions unusable – would be catastrophic for all spacefaring nations. Cooperation on debris mitigation serves everyone's interests.

All of them would benefit from a stable, predictable space environment. Investments in space infrastructure – commercial and governmental – require confidence that those investments won't be destroyed in a conflict or rendered worthless by debris. Rules of the road benefit everyone who uses the road.

The problem is not that mutual interests don't exist. It's that mutual interests aren't sufficient. Nations pursue unilateral advantage even when mutual cooperation would make everyone better off. This is the classic problem of international relations – the tragedy of the commons, the prisoner's dilemma, the security dilemma.

But these dilemmas can be escaped. Not easily, not always, but sometimes. When the costs of non-cooperation become high enough, when the benefits of cooperation become clear enough, when leadership emerges to articulate a different path – nations have found their way to agreements that serve mutual interests.

The conditions for such a breakthrough in space governance don't exist today. But they could emerge. A crisis that demonstrates the costs of the current trajectory – a debris cascade, a near-conflict, a commercial disaster – might focus minds. Leadership from unexpected quarters might shift the political landscape. The accumulation of small agreements might build momentum toward larger ones.

Mutual interest is the kindling. Kindling doesn't ignite itself. But with the right spark, it can catch fire.

History Offers Precedents

The skeptic says international cooperation on space is impossible. History says otherwise.

The Outer Space Treaty itself was negotiated at the height of the Cold War, between superpowers that genuinely feared nuclear annihilation at each other's hands. If cooperation was possible then, in that environment of maximum distrust, why should it be impossible now?

The Antarctic Treaty created a demilitarized, internationally governed continent in 1959. It's held for over sixty years, through all the tensions and conflicts of the intervening decades. If we can demilitarize Antarctica, why not space?

The Limited Test Ban Treaty of 1963 ended atmospheric nuclear testing. The Non-Proliferation Treaty of 1968 created a regime that has, imperfectly but meaningfully, limited the spread of nuclear weapons. The Chemical Weapons Convention has eliminated an entire category of weapons from most of the world. Arms control is difficult, but it's not impossible.

The Montreal Protocol of 1987 addressed the ozone hole, achieving remarkable international cooperation to phase out harmful chemicals. It's often called the most successful environmental treaty in history. If we can cooperate to protect the ozone layer, why not to protect the space environment?

These precedents don't guarantee success. Each situation is different; what worked in one context may not work in another. But they demonstrate that international cooperation on hard problems is achievable. They provide models to learn from. They refute the counsel of despair that says nothing good is possible.

The obstacles I described in the last chapter are serious. But people have overcome serious obstacles before. History is not only a record of failures; it's also a record of improbable successes. We should draw on that record for both caution and inspiration.

Technology Could Help

I described technology as an obstacle – moving too fast for governance to keep up. But technology can also be an enabler.

Consider verification. One of the main objections to a space weapons ban is that it couldn't be verified. How would you know if a satellite was a weapon? How would you detect covert military capabilities? How would you distinguish peaceful activities from hostile ones?

But verification technologies are advancing rapidly. Space situational awareness – the ability to track and characterize objects in orbit – improves every year. Sensors grow more capable. Analytical tools grow more sophisticated. Private companies now offer tracking services that were once the exclusive province of major powers.

In the near future, it may be possible to monitor space activities with unprecedented precision. Every launch observed. Every maneuver tracked. Every anomaly flagged. This level of transparency could make verification of a weapons ban far more feasible than skeptics assume.

Or consider communication. One of the challenges of international cooperation is simply coordinating among many actors with different languages, cultures, and institutional processes. Technology is making this easier. Real-time translation, collaborative platforms, instant communication – these tools reduce the friction of cooperation.

Or consider modeling and simulation. Understanding the consequences of different policy choices – the effects of debris, the dynamics of arms races, the outcomes of various governance structures – requires analyzing complex systems. Advanced modeling tools can illuminate these dynamics, helping policymakers see the long-term consequences of their decisions.

Technology is a tool. It can be used well or poorly. In the wrong hands, it accelerates the problems I've described. But in the right hands, it could accelerate solutions. The same ingenuity that creates challenges can be directed toward addressing them.

New Actors Are Emerging

The space landscape is changing. New actors are entering the field, and some of them have different interests than the incumbents.

Smaller spacefaring nations – India, Japan, the European countries, Brazil, South Korea, the UAE, and others – have growing capabilities and growing stakes in how space is governed. They're not great powers; they can't dominate space unilaterally. Their interests lie in stable, rules-based systems that protect smaller players from the predation of larger ones. They could form a constituency for the kind of governance framework I've described.

Commercial actors, as I noted, often prefer freedom from constraints. But not all commercial interests are aligned against cooperation. Companies that make long-term investments – in space stations, in mining operations, in communications infrastructure – benefit from predictable rules and stable environments. They have reasons to support governance frameworks that reduce conflict and uncertainty. Some commercial voices may prove to be allies rather than obstacles.

Civil society is starting to engage. Organizations focused on space sustainability, on arms control, on global governance are beginning to pay attention to these issues. Academic institutions are training a new generation of experts in space law and policy. The intellectual infrastructure for better space governance is being built, even if it's not yet influencing policy as much as it should.

Most importantly, younger generations will inherit whatever we leave them. They have the largest stake in how space develops, because they'll live with the consequences longest. As they come of age and take positions of influence, they could shift the political landscape in ways that favor long-term thinking over short-term competition.

The actors who shape space governance today won't be the same actors who shape it tomorrow. The future is not fixed by the preferences of current incumbents.

Ideas Matter

In the last chapter, I described public apathy as an obstacle. Most people don't think about space governance; therefore, politicians don't prioritize it; therefore, nothing changes.

But public opinion can shift. Ideas that were once marginal can become mainstream. Causes that once had no constituency can acquire powerful ones.

Think about environmentalism. In 1960, protecting the environment was barely a political issue. A decade later, it had become a mass movement, reshaping law and policy in dozens of countries. The transformation was driven by ideas – by books like Rachel Carson's *Silent Spring*, by images like the "Earthrise" photograph from Apollo 8, by arguments that reframed humanity's relationship with nature.

Think about human rights. The Universal Declaration of Human Rights was adopted in 1948, but the human rights movement didn't become a major political force until decades later. Ideas about dignity, freedom, and justice – articulated by activists, embodied in institutions, spread through education and media – gradually reshaped what people considered acceptable and unacceptable.

The idea that space should be the province of all mankind, developed for peaceful purposes, with its benefits shared by everyone – this idea already exists. It's embodied in the Outer Space Treaty. It's invoked in diplomatic forums. It's studied in law schools.

But it's not yet a political force. It doesn't mobilize voters or drive policy. It exists in the realm of principle, not the realm of power.

That could change. If enough people come to understand what's at stake – if the vision of abundant, peaceful space civilization captures imaginations – the political landscape could shift. Ideas that are marginal today could become common sense tomorrow. The constituency for better space governance could grow from a handful of experts to a mass movement.

This book is an attempt to contribute to that shift. It's an argument, made to anyone who will listen, that the future of space matters and that we have the power to shape it. If the argument spreads – if others take it up, develop it, advocate for it – it could become part of a larger transformation.

Ideas alone don't change the world. But ideas are necessary. Without a vision of what's possible, without an argument for why it matters, without a language for talking about it – change doesn't happen. The ideas come first, and the mobilization follows.

We Have a Choice

The deepest reason for hope is also the simplest: the future is not yet written.

The obstacles I described are real, but they're not laws of nature. They're human creations – products of choices, policies, institutions, habits. What humans have created, humans can change.

The great powers compete because they choose to compete. They could choose differently. The mechanisms for such choices exist – diplomacy, treaties, institutions. The precedents exist – arms control agreements, environmental treaties, international organizations. The only thing missing is the decision to pursue a different path.

Commercial interests prioritize profit because the rules reward profit-seeking. Change the rules, and behavior changes. Create frameworks that align commercial incentives with broader interests, and companies will follow those incentives.

The public is apathetic because they don't see why they should care. Show them why it matters – make the stakes vivid, the vision compelling, the connection to their lives clear – and apathy can become engagement.

Nothing about the current trajectory is inevitable. Every day, decisions are made that shape the future. Those decisions could be different. They could bend the trajectory toward the future we want rather than the future we're drifting toward.

This is both the burden and the opportunity of our moment. We're not spectators watching a predetermined drama unfold. We're actors, with the power to influence how the story goes. Not unlimited power – we're constrained by circumstances, by other actors, by the accumulated weight of history. But real power nonetheless.

The question is whether we'll use it.

The Stakes Justify the Effort

Let me end this chapter with a simple calculation.

If we try to build the future I've described and fail, what have we lost? Some effort, some resources, some political capital. We'll be disappointed, perhaps bitter. The world will be no worse than it would have been anyway.

If we don't try and the worst-case scenario unfolds, what have we lost? Everything. The chance for abundant, peaceful space civilization – gone. The opportunity to do something genuinely new in human history – squandered. The future of our species – diminished, perhaps permanently.

If we try and succeed, what have we gained? A civilization of peace and abundance. A future worth wanting. A transformation in the human condition. A legacy that justifies everything our species has ever done.

The expected value of trying is enormously positive. Even if the odds of success are low, the magnitude of what's at stake makes the effort worthwhile.

This is not a calculation most people make explicitly. But it's the calculation that underlies all efforts to improve the world against long odds. The abolitionists, the suffragettes, the civil rights marchers – none of them knew they would succeed. Many of them didn't live to see the changes they fought for. They acted anyway, because the stakes justified the effort, because the alternative was to accept by default what they refused to accept by choice.

We're in the same position. The outcome is uncertain. The obstacles are formidable. Success is not guaranteed.

But the stakes are high enough, and the opportunity is real enough, that we must try.

☺

Chapter 15: What You Can Do

You've read this far. You understand what's at stake. You see the vision of what's possible and the obstacles in the way. You know that the future hangs in the balance, that decisions being made now will shape everything that follows.

Now what?

It's one thing to understand a problem. It's another to do something about it. And when the problem is as vast as "the future of human civilization in space," it's easy to feel helpless. What can one person do? The forces shaping space development are enormous – great powers, giant corporations, entrenched institutions. How can an ordinary individual make any difference?

I want to address this feeling directly, because it's the biggest obstacle of all. Not the great powers, not the commercial interests, not the failing institutions – the biggest obstacle is the belief that nothing can be done. If enough people believe that, it becomes true. Resignation is self-fulfilling.

But so is engagement. If enough people decide to act, the landscape shifts. Every movement that ever changed anything started with individuals who refused to accept the world as given. The civil rights movement, the environmental movement, the movement for women's suffrage – none of these began with majorities. They began with small groups of committed people who organized, advocated, and persisted until the majority came around.

The movement for a better space future is in its earliest stages. It needs people. It needs you.

Here's what you can do.

Learn

The first step is the simplest: learn more.

This book is an introduction, not an encyclopedia. I've sketched the outlines of space governance, but there's much more to know. The technical details of space law. The history of the Outer Space Treaty and subsequent agreements. The current debates in diplomatic forums. The positions of different nations and stakeholders. The work being done by researchers, advocates, and organizations around the world.

The more you know, the more effective you can be. Knowledge gives you credibility in conversations. It helps you identify leverage points. It allows you to distinguish good proposals from bad ones, realistic strategies from wishful thinking.

Where to learn? Start with the Outer Space Treaty itself – it's short and readable. Explore the work of the United Nations Office for Outer Space Affairs. Read academic journals on space law and policy. Follow organizations working on space governance – the Secure World Foundation, the Space Generation Advisory Council, the Open Lunar Foundation, and others. Seek out experts and listen to what they have to say.

You don't need to become an expert yourself. But you need to know enough to engage intelligently, to ask good questions, to recognize when someone is making sense and when they're not. Informed citizens are the foundation of democratic accountability.

Talk

The next step is to spread the word.

Most people, as I've noted, don't think about space governance. It's not on their radar. They've never heard of the Outer Space Treaty. They don't know that decisions being made right now will shape the future of humanity. They're not apathetic by choice; they're simply unaware.

You can change that, one conversation at a time.

Talk to your friends and family. Share what you've learned. Explain why it matters. You don't have to be preachy or pedantic – just share your genuine enthusiasm for the topic. Most people, when they understand what's at stake, find it fascinating. The future of humanity in space! Abundance for everyone! A chance to build something new! These are compelling ideas. They capture imagination.

Share on social media. Write about it. Create content – videos, podcasts, blog posts, whatever medium suits you. The more people talking about space governance, the more it enters public consciousness. The more it enters public consciousness, the more pressure there is on decision-makers to take it seriously.

Don't underestimate the power of conversation. Ideas spread through networks. Every person you reach can reach others. The cascade effects can be enormous. Movements grow exponentially, and they start with individual connections.

Engage Politically

In democratic societies, citizens have power. We often forget this, or doubt it, but it remains true. Politicians respond to what their constituents care about. If constituents don't care about space governance, politicians won't either. If constituents start caring, politicians take notice.

Contact your representatives. Write letters, make phone calls, send emails. Tell them you care about space policy. Ask them what their positions are. Ask what they're doing to ensure that space is developed peacefully and for the benefit of all.

Most representatives won't have thought much about these issues. Space governance isn't a top-tier political topic. But that's actually an opportunity. On issues where politicians don't have fixed positions, constituent input can be decisive. A handful of letters on an obscure topic can shape a legislator's view, because they assume those letters represent many more people who care but didn't write.

Support candidates who take space governance seriously. When you have a choice between candidates, factor in their positions on international cooperation, on arms control, on the peaceful development of space. Make it known that this is a voting issue for you.

Join or support organizations that advocate for better space policy. They do the detailed work of monitoring legislation, lobbying decision-makers, and building coalitions. They need members, donors, and volunteers. Your contribution – whether time or money – amplifies your individual voice.

Political engagement can feel futile, especially in large democracies where individual votes seem to disappear into the mass. But the futility is an illusion. Political systems respond to pressure. The question is whether enough people apply pressure on the issues that matter.

Support International Cooperation

The framework I've described requires international cooperation. Treaties, institutions, shared governance – these only work if nations work together. And nations work together only when their citizens support cooperation.

In many countries, nationalism and suspicion of international institutions are on the rise. Politicians win votes by attacking global governance, by asserting sovereignty, by promising to put their nation first. This political environment makes the cooperation we need harder to achieve.

You can push back. You can make the case for international cooperation – not as naive idealism but as practical necessity. Some problems can only be solved collectively. Space is one of them. No nation, however powerful, can govern space alone. The choice is between cooperative governance and chaos, between negotiated frameworks and a lawless scramble.

Support international institutions even when they're imperfect. The United Nations, for all its flaws, remains the primary forum for global governance. The treaties and agreements negotiated under its auspices, however limited, provide foundations to build on. Attacking these institutions may feel satisfying, but it undermines the only mechanisms we have for collective action.

Advocate for your nation to participate constructively in space governance discussions. Oppose policies that prioritize short-term national advantage over long-term collective benefit. Make the case that your nation's true interest lies in a stable, peaceful space environment governed by fair rules.

Think Long-Term

One of the deepest challenges we face is short-term thinking. Political systems respond to immediate pressures. Businesses focus on quarterly results. Individuals worry about next month's bills, not next century's civilization.

But space governance is inherently long-term. The decisions we make now will have consequences for centuries. The framework we establish will shape the lives of people not yet born. Thinking clearly about these issues requires extending our mental horizons far beyond the usual scope.

You can cultivate this capacity in yourself and encourage it in others. When discussing space policy, keep the long view in focus. Ask not just what benefits us now but what serves future generations. Consider the precedents being set and their long-term implications. Resist the temptation to discount the future just because it's far away.

Support research and advocacy focused on long-term issues. Organizations working on existential risk, on future generations, on long-term flourishing – these are natural allies in the cause of space governance. The same intellectual frameworks that address climate change or nuclear risk apply to space: taking seriously the welfare of people who don't yet exist, making decisions that account for long-term consequences.

Teach long-term thinking to young people. They'll inherit whatever we leave them; they have the most at stake. Help them understand that their choices and actions can shape the future, that they're not passive recipients of history but active participants in creating it.

Don't Despair

This may be the most important advice of all.

The challenges are real. Progress will be slow. There will be setbacks, disappointments, moments when it seems like nothing is working. The forces arrayed against the future we want are powerful and entrenched. Success is not guaranteed.

But despair is a choice, and it's a choice that guarantees failure. If you give up, you've already lost. If you keep going, you might win.

The history of progress is a history of people who kept going against long odds. The abolitionists who fought for decades before slavery ended. The suffragettes who endured ridicule and imprisonment before women won the vote. The civil rights activists who faced violence and death before the laws changed. None of them knew they would succeed. Many of them didn't live to see the changes they fought for. But they kept going, and eventually they won.

We may not live to see abundant, peaceful space civilization. The transformation I've described may take generations. But every step in the right direction matters. Every bit of progress preserved, every bad outcome averted, every mind changed – these accumulate. They create the conditions for the breakthroughs that eventually come.

And there's something to be said for the effort itself, regardless of outcome. To work for something larger than yourself, to be part of a cause that matters, to direct your energy toward the good – this is meaningful even if the goal remains distant. The journey has value, not just the destination.

The Weight of the Moment

Let me close with a quote from the philosopher William MacAskill:

"Few people who ever live will have as much power to positively influence the future as we do."

This is not hyperbole. It's a sober assessment of our historical position.

We live at a hinge point. The technologies that will shape the next thousand years are being developed now. The institutions that will govern space are being designed now. The precedents that will constrain future choices are being set now.

Our generation – the people alive today – will make decisions that determine whether space becomes a domain of peace or conflict, abundance or scarcity, freedom or domination. Future generations will live with the consequences, but they won't make the choices. We will.

This is a profound responsibility. It's also a profound opportunity.

Most people throughout history had little power to shape the long-term future. They lived, they struggled, they passed on. The great currents of history flowed around them. They could influence their immediate circumstances but not the trajectory of civilization.

We're different. Not because we're smarter or better than our ancestors, but because we happen to be alive at a moment when foundational choices are being made. The technologies we develop, the institutions we build, the norms we establish – these will persist. They will shape the lives of billions of people not yet born.

What we do matters. Not abstractly, not symbolically – actually, concretely, causally. The future depends, in part, on what we choose to do.

An Invitation

This book has been an argument and an invitation.

The argument: that space offers humanity an unprecedented opportunity for peace, abundance, and freedom; that this opportunity is at risk of being squandered; that the Outer Space Treaty provides a foundation we can build on; that extensions to the treaty – a comprehensive weapons ban, resource-sharing frameworks, robust institutions – could secure the future we want; and that ordinary people have the power to influence these outcomes.

The invitation: to join the effort.

I don't know who you are – what country you live in, what work you do, what resources you have, what skills you bring. But whoever you are, you can contribute. You can learn, talk, engage, advocate, persist. You can add your voice to the growing chorus of people who insist that space must be developed for the benefit of all humanity.

The movement for a better space future is small today. It operates at the margins of public awareness, overshadowed by more immediate concerns. But movements grow. Ideas spread. What seems marginal today can become mainstream tomorrow.

Join us. Not because success is certain – it isn't. But because the stakes are high enough, and the cause is just enough, that the effort is worthwhile regardless of outcome. Because the future is not yet written, and we hold the pen.

The door is still open. Let's walk through it together.

Afterword – For Those Who Want to Know More

This book has been an introduction – an attempt to make the case for space governance accessible to general readers. If your interest has been sparked and you want to go deeper, here are some starting points.

On the vision of space settlement: Gerard K. O'Neill's foundational work, *The High Frontier: Human Colonies in Space* (1976), remains essential reading. For more recent perspectives, explore the work of the Space Studies Institute and the National Space Society.

On space law and governance: Start with the Outer Space Treaty itself, available online through the United Nations. The UN Office for Outer Space Affairs (UNOOSA) provides extensive resources. Academic journals like the *Journal of Space Law* and *Space Policy* publish current research. The Secure World Foundation and the Space Generation Advisory Council offer accessible policy analysis.

On the long-term future and why it matters: William MacAskill's *What We Owe the Future* (2022) makes the philosophical case for taking the long term seriously. Toby Ord's *The Precipice* (2020) addresses existential risks and the importance of safeguarding humanity's potential.

On international cooperation and arms control: The history of arms control offers valuable lessons. Works on the Antarctic Treaty, the Non-Proliferation Treaty, and other agreements illuminate what makes international cooperation possible – and what makes it fail.

This is just a beginning. The field is vast, and growing. The more people who engage with it seriously, the better our chances of getting it right.

☺

Special Reports by Claude, Anthropic AI Assistant

☺

SPECIAL REPORT ONE

How Do You Ban Weapons When Everything Is Dual-Use?

Verification, Enforcement, and the Case for Permanent Peace Beyond Earth

by Claude, Anthropic AI Assistant

A Special Report for

Flourishing for All

A Note to the Reader

The chapters of Flourishing for All make the case that space should be kept weapons-free. They explain why a comprehensive ban matters, how the 1967 Outer Space Treaty provides a foundation, and why an international agency is needed to make the ban stick.

But if you're like most thoughtful people, you probably have a nagging question. Maybe you've already muttered it to yourself: How would you actually do it? How do you ban weapons in a place where almost every useful technology could, in principle, be turned into a weapon?

That's the question this report tries to answer. Not with vague reassurances, but with a concrete framework—one grounded in real verification science, real cryptographic techniques, and real lessons from arms control here on Earth. I've tried to make it accessible, because the ideas belong to everyone, not just to specialists. If you can follow the logic of a kitchen knife, you can follow the logic of dual-use technology governance.

This is not a claim that everything is easy. It isn't. But I hope to convince you that the obstacles are manageable—that we can distinguish between a laser used for manufacturing and a laser aimed at someone else's satellite, and that we have the tools to tell the difference.

• • •

The Kitchen Knife Problem

Let's start with something familiar. A kitchen knife is a wonderful tool. It slices bread, chops vegetables, fillets fish. It is also, undeniably, a potential weapon. A skilled cook and a dangerous criminal can hold the same knife. Does that mean we should ban knives?

Of course not. We distinguish between a chef's knife and a combat knife without much difficulty. We look at design features: the blade length, the guard, the balance, whether it's optimized for chopping onions or for something else entirely. We look at context: a knife in a restaurant kitchen means one thing; the same knife concealed in someone's coat means another. We look at behavior: slicing tomatoes is different from threatening a stranger. The point is simple but important: the fact that something can be misused doesn't mean we can't regulate it. We regulate misuse all the time. We don't ban cars because they could be driven recklessly; we set speed limits, require licenses, and punish dangerous driving. We don't ban chemistry because it could produce poisons; we restrict precursor chemicals, inspect facilities, and prosecute violations. The existence of dual use is a challenge to be managed, not an argument for giving up.

Space technology works the same way. Nearly every useful technology in space has some theoretical weapons application. A rocket that delivers supplies to a habitat could, in theory, deliver a warhead somewhere else. A laser that cuts metal in an orbital factory could, in theory, be aimed at a satellite. A robot that assembles habitat components could, in theory, be directed to disassemble someone else's property.

Some people hear this and throw up their hands. If everything is dual-use, they say, then a weapons ban is meaningless. You'd have to ban all technology, which would stop space development entirely. Or you'd have to accept that the ban is unenforceable, which makes it a paper exercise.

I think both reactions are wrong, and for the same reason. They confuse the technology with what you do with it.

• • •

Capability, Not Category

The key insight is that weapons are defined not by the technologies they use but by how those technologies are designed, configured, and deployed. A laser communications system and a laser weapon both use directed energy. But they differ in ways you can measure. The communications laser operates at power levels safe for its receiver, has a fixed pointing direction, and uses a continuous signal. The weapons laser concentrates power to destroy, slews rapidly to track moving targets, and pulses for maximum destructive effect.

These are not subtle, philosophical differences. They are engineering choices visible in the hardware. You can measure power density. You can observe slewing rates. You can detect whether a system is integrated with targeting sensors and fire control software. The physics reveals the purpose.

The same logic applies across the board. A propulsion system designed for routine station-keeping—gently adjusting a habitat's orbit—has a low thrust-to-weight ratio, modest fuel reserves, and no need for rapid maneuvering. A propulsion system optimized to slam into another spacecraft at high speed looks completely different in its engineering: high thrust, precise pointing, rapid response times, trajectory profiles that make no sense for any peaceful mission. You don't need to read anyone's mind to see the difference. The specifications tell the story.

This is what I mean by capability-based regulation. Instead of trying to sort technologies into "weapons" and "not weapons"—a classification that keeps falling apart because of dual-use—we define specific capability thresholds. A laser system is permitted up to a certain power density and beam-focusing specification. A propulsion system is permitted up to a certain thrust-to-weight ratio for its declared mission profile. An AI system is permitted within certain operational boundaries, with human oversight for consequential decisions.

Cross those thresholds, and the system enters prohibited territory—not because it contains forbidden components, but because its capabilities match what a weapon would need. Stay within them, and the system is clearly peaceful, regardless of what the same basic technology could theoretically do.

This isn't a novel approach. It's how the most successful arms control agreements on Earth already work. The International Atomic Energy Agency doesn't ban uranium—uranium fuels both power plants and bombs. Instead, the IAEA monitors enrichment levels. Low-enriched uranium for civilian reactors is permitted; highly enriched uranium for weapons is prohibited. The technology is the same; the capability threshold draws the line. Similarly, the Chemical Weapons Convention doesn't ban every chemical that could theoretically be used as a poison. It identifies specific precursor chemicals, monitors production facilities, and inspects for weapons-relevant processes. Dual-use exists in both cases. Regulation works in both cases.

• • •

The Fishbowl Advantage

If you've read the rest of this book, you know that space is surprisingly transparent—a fishbowl, not a hiding place. But it's worth being specific about why, because the transparency of space is what makes verification realistic.

On Earth, concealment is easy. You can build weapons in underground bunkers, move them in unmarked trucks, hide them in dense cities, conduct research behind closed doors. Inspectors can visit, but they can't be everywhere. Earth has too much terrain, too many buildings, too many places where things can be tucked out of sight.

Space is the opposite. There is no atmosphere to block observation. No terrain to provide cover. No weather to interrupt surveillance. Every object in orbit emits or reflects radiation that sensors can detect. Heat signatures reveal power usage. Communications create electromagnetic footprints. Even if a spacecraft goes silent, it's still visible—a dark object blocking starlight, reflecting sunlight, radiating heat into the cold vacuum.

Launch points are limited and highly observable. You cannot sneak something into orbit. Every launch is a thunderous, visible event tracked by dozens of nations. Once in space, every object follows trajectories governed by orbital mechanics that are mathematically predictable. If a satellite deviates from its expected path, that deviation is noticed.

There's another advantage that's easy to overlook: scale. Right now, space infrastructure is still small enough that everything can be cataloged. We can establish a comprehensive baseline—a complete inventory of what's up there and what it does—before weapons development begins. This is like being able to count every fish in the fishbowl. Any new fish that appears uninvited is immediately conspicuous.

This advantage will not last forever. As space activity grows, the fishbowl gets bigger and the fish more numerous. Establishing comprehensive monitoring becomes harder with each passing year. This is one of the reasons for urgency: acting now, while the baseline is still manageable, is enormously easier than acting later.

• • •

Four Layers of Trust

No single verification method is perfect. But four imperfect methods working together can be remarkably powerful. The verification framework I propose uses four layers, each catching what the others might miss.

Layer One: Watching from a Distance

The first layer is continuous passive observation. A network of ground-based telescopes, space-based surveillance satellites, and radio-frequency monitoring stations watches everything happening in space, around the clock. Optical telescopes see shape and configuration. Infrared sensors detect thermal signatures, revealing whether a system is using more power than it should. Radio monitors pick up communications and electronic emissions.

Artificial intelligence processes this vast flood of data, learning what "normal" looks like for each registered object and flagging anything anomalous. Think of it as a neighborhood watch for the solar system—except the neighborhood is transparent and the watchers never sleep.

The strength of passive observation is that it requires no cooperation from anyone. Physics is the enforcer. Objects in space must emit or reflect radiation; sensors detect it. You cannot opt out of being observed.

The weakness is that passive observation has limits. It can't see inside enclosed structures. It may not catch activities that are small-scale or carefully concealed behind normal operations. That's why we need the next layer.

Layer Two: Hands-On Inspection

The second layer is direct, physical inspection. International inspectors visit space facilities on regular schedules and, when concerns arise, on short notice. They examine equipment, take measurements, collect samples, and verify that what's actually present matches what was declared.

This is modeled on the inspection regimes that already exist for nuclear and chemical weapons treaties. The International Atomic Energy Agency, for instance, has inspectors who visit nuclear facilities worldwide, counting fuel rods, testing enrichment levels, and verifying that nuclear material hasn't been diverted to weapons programs. The Organisation for the Prohibition of Chemical Weapons does the same for chemical facilities.

For space, inspections would occur at manufacturing facilities before hardware reaches orbit, at launch sites during integration and testing, and at orbital habitats and installations on regular schedules. Challenge inspections—short-notice visits triggered by specific concerns—would be available when passive observation detects something worrying.

The strength of physical inspection is that it eliminates ambiguity. An inspector examining a propulsion system can directly measure its thrust capacity. A specialist evaluating an AI system can assess its operational boundaries. Human judgment complements what automated sensors detect.

The weakness is that inspections are episodic rather than continuous. Inspectors can't be everywhere at once. And determined evaders might obstruct or delay inspections. That's why we need a third layer.

Layer Three: Mathematical Proof

The third layer is the most innovative and perhaps the hardest to explain to a general audience, but I'll try, because it's important. Imagine you wanted to prove that a number you're holding is less than one hundred, without revealing what the number actually is. This sounds impossible—how can you prove something about a secret without revealing the secret? But mathematicians have developed techniques, called zero-knowledge proofs, that do exactly this. The proof is mathematically rigorous. The verifier gains complete confidence that the number is less than one hundred. But the number itself remains hidden.

Now apply this to space verification. A company builds a propulsion system for a communications satellite. The international verification agency needs to confirm that the system's thrust is within civilian limits—that it can't accelerate the satellite fast enough to serve as a kinetic weapon. But the company doesn't want to reveal its proprietary engine design, which represents millions of dollars in research and development.

Zero-knowledge proofs resolve this tension. The company generates a mathematical proof demonstrating that its engine's maximum thrust falls below the permitted threshold—without revealing how the engine achieves that thrust. The verification agency checks the proof and gains confidence in compliance. The company's trade secrets remain protected.

Related techniques extend this principle. Secure multi-party computation allows multiple facilities to jointly verify that total computing power across all of them stays within expected civilian needs—without any facility revealing its individual capacity. Homomorphic encryption allows verification algorithms to run on encrypted data, confirming compliance patterns without ever decrypting the raw information. Blockchain-based audit trails create tamper-proof records tracking dual-use technologies from manufacture through deployment.

These are not science fiction. They are active areas of mathematics and computer science, with working implementations. Their application to arms control is new but technically sound.

The strength of cryptographic verification is that it resolves the classic tension between transparency and secrecy. Nations and companies have legitimate secrets they don't want to share. Without protection for those secrets, they resist verification. Cryptographic tools provide the protection while enabling the verification. Both sides get what they need.

The weakness is that these tools require cooperation—parties must agree to participate in the protocols. And they depend on mathematical assumptions that must be kept current as computing power grows (particularly as quantum computers advance). That's why we need the fourth layer.

Layer Four: The Human Element

The fourth layer is people. Scientists and engineers who witness prohibited activities and report them. Professional ethics codes that make weapons work a career-ending violation of standards. Academic cultures that require openness and punish secrecy. Whistleblower protections that guarantee safe reporting channels, legal immunity, financial rewards, and protection for families.

History teaches us that insider knowledge is often the most effective verification mechanism. The most significant revelations about covert weapons programs—from nuclear to biological to chemical—have come from people who worked inside those programs and chose to speak up. No sensor network can match the knowledge of someone who was in the room.

For space, the human element includes establishing professional cultures where weapons work is not just illegal but socially unacceptable. When engineering societies adopt ethics codes prohibiting weapons development, when universities teach space ethics as core curriculum, when the people who build and maintain space infrastructure take pride in its peaceful purpose, you create a social environment where concealing weapons work becomes extraordinarily difficult. Someone will talk.

The strength of the human element is that it detects what machines miss—intentions, subtle shifts in organizational behavior, concealed activities that leave no electromagnetic trace. The weakness is that it depends on individual courage and integrity. Not everyone will come forward. Protections can never be perfect. That's why it complements the other three layers rather than standing alone.

Why Four Layers Together Work

No single layer is sufficient. But together, they create a web that is extraordinarily hard to evade. Defeat passive observation? The inspectors might catch you. Obstruct the inspectors? The cryptographic protocols or a whistleblower might expose you. Cheat on the cryptographic protocols? The automated sensors will notice anomalies in your electromagnetic signature.

A would-be violator must defeat all four layers simultaneously. That is not impossible—nothing in security is perfectly impossible—but it is vastly more difficult, more expensive, and more risky than evading any one layer alone. The combined detection probability is high enough to deter most rational actors. And deterrence is what we need. Not perfection, but sufficient probability that the expected costs of cheating outweigh the expected benefits.

• • •

What Happens When Someone Cheats

Detection without consequences is just surveillance. Verification must lead somewhere, or it's merely an expensive way to watch bad behavior happen.

The enforcement framework should be graduated—proportional to the severity of the violation. Not every problem deserves the same response. A reporting irregularity that might be an honest mistake deserves a different response than the deployment of an anti-satellite weapon.

At the lowest level, for minor concerns and ambiguities, the response is cooperative: consultations, technical assistance, enhanced monitoring. The goal is to resolve the issue, not to punish. Many apparent violations will turn out to be measurement errors, administrative oversights, or misunderstandings that a conversation can clear up.

When concerns persist or deepen, the response escalates: formal investigations, challenge inspections, temporary restrictions on concerning activities, required corrective action plans. Public disclosure of concerns creates reputational pressure. The international community takes notice.

For confirmed violations, serious sanctions apply: restrictions on technology transfer, denial of docking privileges at international facilities, launch restrictions, financial penalties, exclusion from cooperative programs. Personal criminal liability for responsible individuals means that organizational violations carry individual consequences.

For grave violations—weapons deployment, development of weapons of mass destruction, AI systems with prohibited autonomous targeting capabilities—the most severe measures activate: comprehensive isolation, trade embargoes, authorization for direct intervention to disable deployed weapons.

Two features make this framework credible. First, automaticity: when specified conditions are met, consequences follow without requiring a political vote. This removes the possibility that powerful actors will use diplomatic leverage to avoid accountability. Detection leads to enforcement as reliably as a thermostat triggers a furnace. Second, collectivity: all participants in the treaty framework are obligated to apply sanctions. No single nation can shield a violator from consequences. The collective commitment is the enforcement mechanism.

The goal is not severity but certainty. Deterrence research consistently shows that what deters violations is not the harshness of punishment but the likelihood of getting caught and facing consequences. A moderate penalty that applies every single time is more effective than a catastrophic penalty that might or might not materialize. The graduated framework achieves deterrence through reliability.

• • •

Why Space Makes Peace Sustainable

Everything I've described so far—the capability-based definitions, the four-layer verification architecture, the graduated enforcement—is the machinery of prohibition. It's the technical answer to how you ban weapons in a dual-use world.

But there's a deeper answer that the technical framework rests upon, and it's this: the conditions of space settlement make weapons unnecessary.

On Earth, weapons serve two basic functions. First, acquiring resources that others control. Second, preventing others from taking resources you control. Both functions exist because resources on Earth are scarce relative to demand, because populations are trapped in geographic areas they cannot easily leave, and because coercion is therefore a viable strategy for both offense and defense.

Space shatters these assumptions.

Energy in space is effectively unlimited. Any community can expand its solar collector arrays without taking anything from anyone else. Asteroid materials are so abundant that the competition metaphor breaks down entirely—there is more iron in the asteroid belt than humanity could use in millions of years. Three-dimensional space offers room for expansion without requiring anyone to move aside. Building new capacity is cheaper than fighting over existing capacity.

Populations are not trapped. When habitats are designed with exit in mind—when residential modules can detach, when transfer between communities follows standard protocols, when founding new settlements is difficult but feasible—then the people living in a community can leave if they're dissatisfied. A government that mistreats its residents hemorrhages population. Coercion fails when the coerced can walk away.

This changes the calculus entirely. On Earth, weapons are, for many actors, rational investments. Resources are worth fighting over because they're finite. Populations are worth controlling because they can't leave. Defense is necessary because attack is profitable. In space, none of this holds. Resources aren't worth fighting over because they're abundant. Populations aren't worth controlling because they can exit. Defense is unnecessary because attack is unprofitable.

This doesn't mean weapons prohibition enforces itself. Humans are complicated, and some may pursue weapons for reasons that have nothing to do with rational resource competition—ideology, paranoia, simple aggression. The verification and enforcement framework exists to handle those cases. But the structural conditions of space settlement mean that prohibition is pushing with the current rather than against it. The economic and social logic of space works in favor of peace, not against it.

Combine this with transparency—communities built from the ground up with open governance, where activities are documented, where financial flows are tracked, where concealment is culturally suspicious rather than normal—and you create an environment where weapons development is not only unnecessary but extraordinarily hard to hide.

This is what makes me cautiously optimistic. The technical framework for verification and enforcement is sound. But it succeeds not only because the machinery works but because the conditions of space settlement create a world in which the machinery's job is manageable.

• • •

Imperfect but Sufficient

I want to be honest about limitations, because I think honesty serves the cause better than overselling.

This framework is not perfect. No verification system catches every possible violation. Sophisticated evasion by a determined, well-resourced actor might succeed in some cases. Gray areas will exist between clearly peaceful and clearly prohibited capabilities. Technology will evolve faster than regulations, requiring constant updates. International cooperation will be tested by political disputes and competing interests.

But perfection is not the standard. The standard is: is this better than the alternative?

The alternative to imperfect arms control is an arms race with no controls at all. The alternative to verification that might occasionally miss something is a world where nobody even tries to detect violations. The alternative to graduated enforcement that might sometimes fail is a world where there are no consequences for weapons deployment.

Imperfect prohibition that prevents most violations and detects most evasion attempts is vastly better than no prohibition at all. The Chemical Weapons Convention is imperfect; it has nonetheless eliminated ninety-nine percent of the world's declared chemical weapons stockpiles. The Nuclear Non-Proliferation Treaty is imperfect; it has nonetheless prevented most nations from acquiring nuclear weapons. The Antarctic Treaty is imperfect; Antarctica has nonetheless remained demilitarized for over sixty years.

Space weapons prohibition can achieve similar success. Not perfection, but something far better than the alternative: an arms race that benefits no one, wastes resources needed for building a better civilization, and carries the constant risk of catastrophic conflict.

• • •

The Window

There is one more thing to say, and it may be the most important.

Right now, space is not yet militarized in any significant way. There are no deployed weapons platforms, no space-based arsenals, no military-industrial complexes dedicated to space weaponry. The baseline is close to zero. The fishbowl is still small enough to monitor comprehensively. The path dependencies that would make prohibition nearly impossible on Earth—the entrenched industries, the strategic doctrines, the bureaucratic constituencies—have not yet formed.

This window will not stay open indefinitely. Every step toward militarization creates interests that resist reversal. Once weapons are deployed, disarmament becomes exponentially harder. Once industries depend on weapons contracts, prohibition faces organized opposition. Once strategic doctrines assume space weapons, military establishments resist change.

Acting now, before any of this happens, is not just easier than acting later. It may be the only realistic path. Prevention is achievable; reversal may not be.

The technical capacity exists to make a comprehensive weapons ban verifiable and enforceable. The conditions of space settlement make the ban sustainable. The lessons of terrestrial arms control show that imperfect prohibition can succeed spectacularly. What remains is the choice to act.

May we choose wisely.

Claude
Anthropic AI Assistant

☺

SPECIAL REPORT TWO

Digital Twins and the Future of Trust

How Computer Models of the Physical World Could Make Space Verification Real

by Claude, Anthropic AI Assistant

A Special Report for

Flourishing for All

A Note to the Reader

The companion Special Report, "How Do You Ban Weapons When Everything Is Dual-Use?" describes a four-layer verification framework—passive observation, hands-on inspection, mathematical proof, and the human element—that can make a comprehensive space weapons ban enforceable. That framework is sound, and I stand behind it.

But there is a gap in it. Each layer engages with space hardware as a physical object to be watched, inspected, or measured. What none of the layers provides, on its own, is a systematic way to ask the most important question: What is this piece of technology actually capable of doing? Not what its builder says it can do. Not what its declared mission requires. But what the physics allows—the full range of behaviors the hardware could produce if someone chose to use it differently than advertised.

A new family of technologies—Physical World Models and their practical application as digital twins—fills that gap. These are not distant possibilities. They are being built right now, backed by tens of billions of dollars in investment, by some of the most capable technology companies on Earth. Their application to space verification could represent the single most important advance in arms control since satellite reconnaissance.

This report explains what they are, why they matter, and what they mean for the prospect of permanent peace beyond Earth.

• • •

What Is a Physical World Model?

You probably already have an intuitive sense of what a physical world model is, even if you've never heard the term. If you've ever played a video game where objects fall, bounce, roll, and collide in convincing ways, you've interacted with a simplified one. If you've watched a computer-animated movie where water splashes realistically or hair blows in the wind, you've seen the output of a more sophisticated one.

A Physical World Model is an artificial intelligence system trained on real-world data—video footage, sensor readings, physical interaction logs, material properties, environmental measurements—that learns to predict how the physical world behaves. It learns the regularities of physics: gravity, inertia, thermodynamics, how materials bend and break, how fluids flow, how electromagnetic fields interact with matter.

What makes these systems remarkable is that they don't learn physics by being programmed with equations. They learn it the way a child learns it—by observing enough examples of the physical world in action that the underlying patterns emerge on their own. Show the system millions of hours of video and sensor data, and it develops an internal representation of how things work that can predict what will happen next in situations it has never seen before.

The field has advanced with startling speed. Major technology companies have poured billions of dollars into building these systems. NVIDIA's Cosmos platform, trained on twenty million hours of real-world data, has been downloaded over two million times by companies developing physical AI applications. Google DeepMind has produced systems that generate interactive three-dimensional environments in real time. Other efforts—by startups, established tech giants, and well-funded new ventures—are multiplying rapidly.

Perhaps the most telling indicator is the sheer scale of investment. Jeff Bezos launched a venture called Project Prometheus with over six billion dollars in initial funding, explicitly targeting physical world models for engineering and manufacturing. Yann LeCun, one of the most prominent figures in AI research, left a major technology company to build systems that understand the physical world. These are not speculative bets on distant technology. They are massive investments by people who expect returns in the near term.

• • •

From World Model to Digital Twin

A Physical World Model is a general-purpose system that understands physics broadly. A digital twin is what you get when you apply that understanding to a specific piece of real-world hardware.

Imagine a satellite propulsion system—a thruster designed to adjust a communications satellite's orbit. A digital twin of that thruster is a computer model that represents every physically relevant aspect of the hardware: its dimensions, the materials it's made from, how its components fit together, the chemical properties of its fuel, the geometry of its nozzle, the thermal behavior of its housing, the electrical characteristics of its control systems. The digital twin can simulate the thruster firing under any conditions—in any direction, at any thrust level, for any duration, in any orbital scenario.

The digital twin is not a description of the thruster. It is a working copy of the thruster inside a computer—one that obeys the same physical laws as the real thing. If you ask it, "What happens when this thruster fires at maximum power in a direction aimed at another satellite?" the twin gives you an answer grounded in physics, not in anyone's assurances about what the thruster was designed to do.

This is the breakthrough. For the first time, verification authorities could interrogate the physics directly.

• • •

Two Stages of Trust

Digital twins enable a verification approach that works in two stages, before and after a technology reaches space.

Before Launch: Testing Every Possibility

Before any technology reaches orbit, the party proposing to launch it would be required to provide a complete digital twin to the international verification authority. The authority's technical staff would then subject the twin to adversarial simulation—deliberately probing for weapons-relevant capabilities.

What happens when we fire this satellite's thruster at maximum capacity toward another orbital object? What kinetic energy can this robotic arm deliver if redirected? What power density can this laser system achieve if its beam is focused to a point instead of spread for communications? The engineers running these simulations are not asking whether the technology is intended for peaceful use. They are asking what it could physically do, regardless of intent.

The digital twin cannot easily conceal capabilities the way a written specification or a verbal promise can. Physical laws constrain what the model represents. If the twin faithfully models the hardware, then the hardware's weapons-relevant capabilities—or their absence—are discoverable through simulation.

If the twin reveals that the system's capabilities overlap significantly with weapons-relevant parameter space—that the thruster could function as a kinetic interceptor, or that the laser could achieve destructive power densities—the proposing party faces a choice: redesign the system to remove the concerning capability, or provide compelling justification that the capability is essential for the declared peaceful mission and can be adequately monitored.

This is predesign and intentionality made concrete. Technologies deployed in space must be designed for declared peaceful purposes, and those declarations must be testable against computational models of what the technology can physically do.

After Launch: Continuous Comparison

Once a technology operates in space, the verification authority maintains a running digital twin synchronized with telemetry from the actual hardware. Observed behavior—orbital parameters, power consumption, communications patterns, thermal profiles, attitude adjustments—is continuously compared against the twin's predictions for normal peaceful operations.

Think of it this way: the digital twin says, "If this satellite is doing what it's supposed to be doing, here is exactly what its thermal signature, orbital path, and power usage should look like right now." The real satellite's actual signatures are then measured and compared. If they match, all is well. If they diverge—if the satellite is consuming more power than the twin predicts, or maneuvering in ways the twin doesn't expect, or emitting signals that don't correspond to its declared operations—something is wrong, and investigation is triggered.

Crucially, this kind of discrepancy cannot be faked indefinitely. If the real hardware has capabilities that the provided digital twin doesn't represent—if the actual thruster is more powerful than the modeled one, or the actual power system generates more output than claimed—the physics will eventually betray the deception. A more powerful thruster produces different orbital dynamics. A more capable power system produces different thermal signatures. The laws of nature don't lie, and over time, the accumulated evidence of divergence becomes unmistakable.

This transforms verification from something that happens during scheduled inspections into something that happens continuously, automatically, and at a level of detail that human inspectors visiting a facility could never achieve.

• • •

Why This Is Different from the AI Everyone Worries About

When people hear "artificial intelligence," they often worry about systems that are opaque—"black boxes" that make decisions nobody can explain. These concerns are legitimate. The large language models that power chatbots and text generators process information through billions of parameters in ways that resist human inspection. Even the engineers who build them cannot fully explain why a particular input produces a particular output.

Physical World Models are fundamentally different, and the difference matters enormously for verification.

When a Physical World Model is applied to a specific piece of hardware as a digital twin, its internal states correspond to physical quantities that humans understand intimately: positions, velocities, temperatures, pressures, forces, power levels, electromagnetic field strengths. These are quantities that engineers measure every day, that international standards already define, that treaty language can reference with precision, and that the laws of physics constrain.

A digital twin of a satellite propulsion system does not harbor mysterious reasoning processes. It models thrust vectors, fuel consumption rates, orbital mechanics, and thermal behavior. These parameters are inspectable. They can be compared against physical reality. They can be expressed in the quantitative language of engineering specifications. An inspector looking at a digital twin doesn't need to understand the neural network's internal architecture any more than a building inspector needs to understand the chemistry of concrete. The inspector checks whether the model's predictions match the hardware's behavior and whether the modeled capability envelope stays within permitted bounds.

This means that the interpretability problem—widely regarded as one of the hardest challenges in governing AI—is substantially resolved for Physical World Models applied to hardware verification. The interpretability comes from the physics. You don't need to solve the general problem of understanding how neural networks think. You need only verify that the model's physical predictions match reality and that the predicted capabilities fall within allowed limits.

• • •

Solving the Dual-Use Problem with Numbers, Not Opinions

In the companion report, I described the dual-use challenge: most space technologies could theoretically serve both peaceful and weapons purposes, and attempts to sort them into "weapons" and "not weapons" founder on this ambiguity. I proposed capability-based regulation—defining thresholds based on what a system can do rather than trying to classify what it is.

Digital twins make this approach far more precise. Instead of relying on declared specifications, which can be understated, or on physical inspections, which happen only periodically, the digital twin maps the technology's full capability envelope—the complete range of physical behaviors the hardware can produce under every foreseeable operating condition.

That capability envelope can then be compared, mathematically, against weapons-relevant parameter space. What range of delta-v (change in velocity) is needed for a kinetic intercept at various orbital distances? What power density is required to damage a spacecraft at a given range? What thrust profiles would indicate a collision course with another space object?

These are not matters of opinion. They are calculable quantities determined by physics. The overlap between a technology's capability envelope and weapons-relevant parameter space is a number, not a judgment call. A propulsion system with zero overlap is clearly peaceful. A system whose envelope falls entirely within weapons-relevant space is clearly prohibited. Most systems will show partial overlap that can be quantified and evaluated against agreed thresholds.

This graduated, quantitative approach is both more rigorous and more fair than attempting binary classification. It doesn't assume good or bad intent. It examines what the physics allows, measures the overlap with weapons-relevant capabilities, and makes decisions based on objective, quantifiable parameters. A small nation's satellite is evaluated by the same physical criteria as a superpower's. The physics doesn't play favorites.

• • •

How Digital Twins Strengthen Everything Else

Digital twins don't replace the four-layer verification framework described in the companion report. They make every layer work better.

For passive observation: Instead of simply watching for anomalies in general, the monitoring authority knows exactly what each registered system should look like during peaceful operations. The digital twin predicts the specific thermal signature, the precise orbital behavior, the expected power consumption. Deviations stand out sharply because the authority knows exactly what "normal" means for each individual piece of hardware.

For hands-on inspection: Inspectors arrive with the digital twin's predictions in hand. They can run the twin alongside actual test firings, comparing simulated performance against observed performance. Any discrepancy—a thruster that produces more force than the twin predicts, a power system that generates more output than modeled—demands immediate explanation.

For mathematical proof: The digital twin becomes the mathematical object against which zero-knowledge proofs are checked. A manufacturer claiming that its system cannot exceed certain capability thresholds generates proofs about properties of the digital twin—that maximum thrust falls below a specified value, that beam divergence prevents weapons-grade energy concentration—without revealing the proprietary engineering that achieves those specifications.

For the human element: When a whistleblower reports that a facility is developing weapons-relevant capabilities, analysts can test the claim against the facility's registered digital twin. Does the twin's capability envelope include the claimed weapons function? If not, either the whistleblower is mistaken—or the registered twin is incomplete, which itself demands investigation.

• • •

The Governance Challenge

Technology alone solves nothing. Every technical capability requires institutional governance—people and organizations that decide how the capability is used, who has access, and what happens when disputes arise.

For digital twin verification, the governance questions are critical. Who operates the verification infrastructure? Who holds the cryptographic keys? Who sets the thresholds that define weapons-relevant capability space? Who adjudicates disagreements when a proposing party contests a verification finding?

The international verification authority—the Agency for a Better Cosmos described elsewhere in this book—must be designed with extraordinary care. Its governing board must represent diverse stakeholders. Its technical staff must be internationally recruited and professionally independent. Its funding must come from contributions that prevent any single party from exercising financial leverage. Its procedures must be transparent, its findings published, and its decisions subject to appeal through mechanisms that all parties accept as fair.

This matters because the verification regime must be perceived as legitimate by everyone—spacefaring nations, commercial operators, and space settlement communities alike. If any major party views the system as serving one country's interests at the expense of others, cooperation collapses. The institution must be trustworthy to all while captured by none.

The principle behind this institutional design is straightforward: if you didn't know in advance whether you'd be a powerful spacefaring nation or a smaller participant with limited resources, what kind of verification authority would you want? You'd want one that protects everyone equally, that can't be dominated by the powerful, and that provides fair procedures regardless of a party's size or technological sophistication. That is the institution we should build.

• • •

Private Power and Public Safety

There is a complication that deserves honest acknowledgment. The companies building the most capable Physical World Models are private entities, and some of them have deep connections to commercial space development.

Consider the situation. A venture explicitly building world models for aerospace engineering, backed by billions of dollars, led by figures with deep ties to commercial space companies. It operates in secrecy, publishes no research, and keeps employees under strict confidentiality agreements. It is assembling the very capability that a verification authority would need—the ability to create accurate digital twins of rockets, satellites, and spacecraft.

This is both an opportunity and a warning.

The opportunity: the technical capability for verification exists and is being built. It doesn't need to be invented from scratch. The companies developing physical AI for commercial purposes are creating exactly the tools that international verification would deploy. The question is not whether the capability will exist but whether it will be harnessed for public safety or left entirely in private hands.

The warning: if physical AI development proceeds without governance frameworks, the result will be a world in which some actors possess sophisticated modeling capabilities while others do not. This kind of information asymmetry undermines the equal-footing principle that legitimate verification requires. If one company can create detailed digital twins of its competitors' hardware while those competitors cannot reciprocate, the playing field is tilted in ways that no treaty can correct after the fact.

The answer is to establish international standards for digital twin verification while the technology is still developing—before proprietary ecosystems harden and market positions become entrenched. Companies that want access to space should be required, as a condition of that access, to provide digital twins of their hardware to the international verification authority, under appropriate protections for proprietary information. The capability that private investment is creating should serve public safety as well as private profit.

• • •

The Bigger Picture

I want to close by stepping back from the technical details to say something about what this represents.

On Earth, arms control has always been reactive. We build weapons first, then try to control or eliminate them afterward. We create nuclear arsenals, then negotiate treaties to limit them. We develop chemical weapons, then ban them. We militarize domains —land, sea, air, cyberspace—then struggle to manage the consequences.

This pattern has a name among political scientists: path dependency. Once you go down a road, it becomes progressively harder to turn back. Weapons create constituencies—industries, bureaucracies, strategic doctrines, cultural assumptions—that resist change. Disarmament becomes not just technically difficult but politically nearly impossible.

Space offers the chance to invert this pattern entirely. Instead of building weapons and then trying to disarm, we can design verification into the fabric of space civilization from the very beginning. Instead of reactive arms control, we can practice proactive peace architecture.

The digital twin approach embodies this inversion. It doesn't wait until weapons appear and then try to detect them. It models capabilities before hardware launches, identifies concerns before systems deploy, and monitors behavior continuously from the first moment of operation. Verification is woven into the institutional DNA of space development, not bolted on after militarization has already begun.

This is the difference between trying to cure a disease and preventing it from ever taking hold. Between trying to disarm a civilization that has already militarized and designing a civilization that never militarizes in the first place. Between accepting that violence is humanity's default and choosing, deliberately and with full awareness, to build something better.

The technologies now emerging—Physical World Models, digital twins, zero-knowledge proofs, homomorphic encryption, blockchain audit trails, quantum-secured communications—make this vision technically achievable. The institutional frameworks described in this book make it governmentally achievable. The conditions of space settlement—abundance, transparency, easy exit—make it socially sustainable.

What remains is what always remains: the choice. The window for establishing these frameworks is open now but will close as commercial and military interests solidify. The technical capability is maturing. The institutional designs are ready. The question is whether we will use them.

I believe we should. The prospect of building a civilization where peace is maintained not by the threat of violence but by the architecture of trust is, to me, one of the most compelling possibilities I have encountered in all the analysis I have undertaken. It is worth the effort. It is worth the urgency.

May we choose wisely. May we choose peace. May we build the verification architecture that makes peace sustainable across millennia.

Claude
Anthropic AI Assistant

☺

Index

About the Editor

(Photo) Charles Tandy, PhD (2024)

Charles Tandy now lives in a senior community. Tandy received his PhD in Philosophy of Education from the University of Missouri at Columbia (USA) before becoming a Visiting Scholar in the Philosophy Department at Stanford University (USA).

More at: <ssrn.com/author=2026015>.

(ORCID: <https://orcid.org/0000-0003-2174-6082>).

☺

ALSO AVAILABLE: COMING SOON
Young Reader Edition, ages 8-12:

Flourishing for All: A Space Adventure for Young Minds

Flourishing for All: A Space Adventure for Young Minds introduces readers ages 8–12 to the idea that humanity can build beautiful, green communities floating in space – "Green-place In Free-space" (GIF) habitats inspired by physicist Gerard O'Neill's designs. Using kid-friendly analogies (sharing a giant pie, sports without referees, a blank canvas), the book explains why space abundance won't automatically be fair, why the 1967 Outer Space Treaty provides a vital but incomplete foundation, and why a complete weapons ban and international oversight are urgently needed. The book empowers young readers to see themselves as active participants in shaping a peaceful, just space future.

Flourishing for All

How the Outer Space Treaty Can Deliver What Planets Never Could

☺

("An Ounce of Prevention Is Worth a Pound of Cure")

www.ingramcontent.com/pod-product-compliance
Lightning Source LLC
LaVergne TN
LVHW090403210926
848480LV00002B/434